Pregnancy and the Pharmaceutical Industry

Pregnancy and the Pharmaceutical Industry

The Movement Towards Evidence-Based
Pharmacotherapy for Pregnant Women

Kristine E. Shields, MSN, DrPH
Women's Health Clinician, Researcher, and Advisor
Shields' Medical Writing and Consulting

ACADEMIC PRESS

An imprint of Elsevier

Academic Press is an imprint of Elsevier
125 London Wall, London EC2Y 5AS, United Kingdom
525 B Street, Suite 1650, San Diego, CA 92101, United States
50 Hampshire Street, 5th Floor, Cambridge, MA 02139, United States
The Boulevard, Langford Lane, Kidlington, Oxford OX5 1GB, United Kingdom

Notices
Knowledge and best practice in this field are constantly changing. As new research and experience broaden our understanding, changes in research methods, professional practices, or medical treatment may become necessary.

Practitioners and researchers must always rely on their own experience and knowledge in evaluating and using any information, methods, compounds, or experiments described herein. In using such information or methods they should be mindful of their own safety and the safety of others, including parties for whom they have a professional responsibility.

To the fullest extent of the law, neither the Publisher nor the authors, contributors, or editors, assume any liability for any injury and/or damage to persons or property as a matter of products liability, negligence or otherwise, or from any use or operation of any methods, products, instructions, or ideas contained in the material herein.

British Library Cataloguing-in-Publication Data
A catalogue record for this book is available from the British Library

Library of Congress Cataloging-in-Publication Data
A catalog record for this book is available from the Library of Congress

ISBN: 978-0-12-818550-6

For Information on all Academic Press publications
visit our website at https://www.elsevier.com/books-and-journals

Publisher: Andre G. Wolff
Acquisition Editor: Erin Hill-Parks
Editorial Project Manager: Sara Pianavilla
Production Project Manager: Poulouse Joseph
Cover Designer: Victoria Pearson

Typeset by MPS Limited, Chennai, India

Working together
to grow libraries in
developing countries

www.elsevier.com • www.bookaid.org

Dedication

This book is dedicated to the pregnant, want-to-be pregnant, and once-pregnant women who encounter(ed) medical or psychological illness.

Also to the three babies I carried who have taught me so much about pregnancy, motherhood, and life.

Contents

Part I
The background, the debate, and the ethics involved

Part II
Quantitative and qualitative discoveries

List of figures

List of tables

Preface

In October 2010, I participated on the organizing committee for the Drug Information Association's Maternal and Pediatric Drug Safety Conference held in Bethesda, MD.[1] One of the invited speakers, Anne Drapkin Lyerly, presented a session entitled "How and When Should Pregnant Women Be Allowed to Participate in Clinical Trials?" She divulged that the pharmaceutical industry had not been included in recent meetings of thought leaders on this topic. She suggested that it was time for the industry to be brought into the dialogue.

The pharmaceutical industry is uniquely positioned to play a meaningful role in this debate because of its influence on the conduct of clinical research. In the United States the industry funds more studies than any other organization including the National Institutes of Health.[2] It is responsible for the research design, the research protocol, the inclusion and exclusion criteria, and the conduct of each of the studies they sponsor. Therefore the pharmaceutical industry has a major influence on the extent to which pregnant women are included in clinical research studies.

In my practice as an OB/GYN Nurse Practitioner, I cared for pregnant women in need of medical intervention. In my role as the director of a pharmaceutical company's pregnancy registry program, I worked in the industry and championed the collection and communication of safety information about drug exposures during pregnancy. I researched, wrote, and presented widely on the topic. With this background in obstetrical practice and pregnancy research within the pharmaceutical industry, I felt that I was uniquely positioned to raise and discuss the issue of pregnant women in clinical trials with my industry colleagues. Knowing health care providers and researchers in industry, at FDA, in medical practice, and within academia, I would be able to access and articulate the issues involved from multiple perspectives. By raising the issue with industry colleagues, and by presenting my findings,

---

1. Drug Information Association (DIA). (October 13–14, 2010). *Maternal and pediatric drug safety conference, Bethesda, MD.*
2. Pharmaceutical Research and Manufacturers of America. (2010). *Press release: R&D investment by U.S. biopharmaceutical companies reached record levels in 2010.* Washington, DC. From <http://www.phrma.org/media/releases/rd-investment-us-biopharmaceutical-companies-reached-record-levels-2010> Retrieved 15.03.11.

I hoped to facilitate the dialogue and encourage their participation in the debate.

I agree with the efforts being made to broaden pregnant women's access to clinical research and believe that progress will not be made without the participation of the pharmaceutical industry. I hope that this book will contribute to these efforts by promoting multidisciplinary dialogue. The ultimate goal is to improve evidence-based medicine for women and their babies.

I want to thank Dr. Lyerly for specifically calling out the pharmaceutical industry as a nonparticipatory stakeholder. She piqued my interest in the topic and it became the subject of my doctoral dissertation and subsequently, this book. She also became a dissertation committee member and provided me with access to her knowledge, her network, and other resources, including her encouragement and support.

I also want to thank Robert G. Sharrar, my mentor and friend, who hired me to run the pregnancy registry program in the pharmaceutical industry. He harnessed my passion for women's health and shaped my career. I can never thank him enough.

Note to the reader: if you are not familiar with industry jargon, pharma is short for pharmaceutical and bio for biotechnology. When I write, I include the biotechnology industry under the pharma umbrella, and what I say about drugs, I mean about vaccines as well. Clinical studies, trials, and research are used interchangeably as are Phases 1, 2, 3, 4 and I, II, III, IV. A "sponsor" is the pharmaceutical company that is responsible for conducting (sponsoring) a clinical trial.

Epigraph

Go out and listen to people, gather the relevant data, frame the issues, pose the questions, and bring everyone, or at least their representatives, to the table...

Barry S. Levy,
1997 Presidential Address to the American Public Health
Association

List of abbreviations

45 CFR 46 Subpart B **(The Common Rule)**	Code of Federal Regulations addressing Additional Protections for Pregnant Women, Human Fetuses and Neonates Involved in Research
AAP	American Association of Pediatricians
ACE	angiotensin converting enzyme
ACOG	American College of Obstetricians and Gynecologists
AIDS	acquired immunodeficiency syndrome
APHA	American Public Health Association
Biotech	biotech industry
BIO	Biotechnology Innovation Organization
CDC	US Centers for Disease Control and Prevention
CEO	chief executive officer
CFR	Code of Federal Regulations
CIOMS	Council for International Organizations for Medical Sciences
COPD	chronic obstructive pulmonary disease
CR	corporate responsibility
CRO	contract research organization
DES	diethylstilbestrol
DSMB	Data Safety Monitoring Board
EMA	European Medicines Agency
EU	European Union
FDA	US Food and Drug Administration
FDAMA	Food and Drug Modernization Act
GBS	group B streptococcus
HHS	(US Department of) Health and Human Services
HIV	human immunodeficiency virus
IOM	Institute of Medicine
IPA	International Pharmaceutical Abstracts
IRB	Institutional Review Board
ISP	International Society of Pharmacoepidemiology
NDAs	new drug applications
NIH	National Institutes of Health
NVP	nausea and vomiting of pregnancy
OB/GYN	obstetrics and gynecology
OPRU	Obstetric-Fetal Pharmacology Research Unit
PD	pharmacodynamics
PDUFA	Prescription Drug User Fee Act
Pharma	pharmaceutical industry

PhRMA	Pharmaceutical Research and Manufacturers' Association
PI	principal investigator
PK	pharmacokinetics
PRGLAC	HHS Task Force on Research Specific to Pregnant Women and Lactating Women
RSV	respiratory syncytial virus
SPH	school of Public Health
UNAIDS	Joint United Nations Programme on HIV/AIDS
UTMB	University of Texas Medical Branch
VICP	Vaccine Injury Compensation Program
WHO	World Health Organization

Part I

The background, the debate, and the ethics involved

Chapter 1

Drug testing and pregnant women: background and significance

The lack of drug studies in pregnancy constitutes a major public health problem.[1]

Background of the issue

"Each year over 400,000 women in the US confront significant medical illness while pregnant."[2] In addition to pregnancy-specific complications like gestational diabetes and preterm labor, medical conditions that occur in non-pregnant women occur in pregnant ones as well, including psychiatric illness, cancer, and infectious diseases. These conditions can have a devastating impact on the health of the pregnant woman and on the well-being of her fetus. The safe and effective treatment of medical conditions complicating pregnancy is challenged by a serious lack of information on the safety and effectiveness of the medications used to treat them. Women's health care practitioners lament that the "current evidence base for the care of pregnant women facing illness is widely regarded as deplorable."[3]

The most effective way to improve our knowledge of safe and effective pharmacotherapy during pregnancy is through clinical research. Yet, despite the recommendations of experts in the US, including the Food and Drug Administration (FDA),[4] the Institute of Medicine (IOM),[5] the Council for International Organizations of Medical Sciences (CIOMS),[6] and the American College of Obstetricians and Gynecologists (ACOG),[7] to include pregnant women in drug research studies, exclusion is the norm. Without information from research studies, clinicians and their pregnant patients must make treatment decisions based on traditions, educated guesses, and gut feelings. Is this really the best we can do?

In 1993, Dr. David Kessler, the FDA Commissioner at the time, stated that "many drugs are ultimately used during pregnancy without reliable data on their maternal and fetal effects." He cited the recent discovery that angiotensin converting enzyme (ACE) inhibitors, first marketed 10 years earlier to

Pregnancy and the Pharmaceutical Industry. DOI: https://doi.org/10.1016/B978-0-12-818550-6.00001-5

treat hypertension, could cause fatal kidney defects in babies when used by pregnant women in their second and third trimester.[8] FDA, he said, needed "to develop recommendations on this important topic that will facilitate the conduct of trials in pregnant women and result in more such trials."[9]

In April 2018, 25 years after the Commissioner first spoke of the need for drug studies in pregnant women, FDA finally released a draft guidance document on the topic: Pregnant Women: Scientific and Ethical Considerations for Inclusion in Clinical Trials.[10]

Historical perspective

As with research practices in general, guidelines regarding women's participation in research have evolved over time. Following the tragic outcomes related to the use of under-studied but regulatory agency-approved products by women in the 1950s and 1960s, particularly thalidomide and diethylstilbestrol (DES), federal regulations were changed to exclude women of childbearing potential from clinical trials.

Thalidomide was never FDA-approved in the US due to safety concerns, but was widely prescribed and sold over-the-counter as a sedative and antiemetic in West Germany (where it was first licensed), Canada, Australia— over 40 countries in total. It was used by pregnant women to prevent and treat nausea and vomiting. In 1962, two physicians, one in Germany and one in Australia, publicized their suspicions that thalidomide was the causative agent in over 10,000 babies born with limb and other organ malformations. Many other thousands of babies died before the drug was removed from the market. Thalidomide has since been reapproved for the treatment of erythema nodosum leprosum (leprosy), multiple myeloma, and other serious conditions and is prescribed in the US via a strict safety program to prevent inadvertent use by pregnant women.

Subsequent to the thalidomide disaster, FDA revamped their regulatory processes, mandating efficacy as well as safety data prior to approval, strengthening regulations to include developmental toxicity testing in at least two species of animals,[11] expanded patient informed consent procedures, and demanded increased transparency from drug manufacturers.[12]

DES was a synthetic estrogen approved by FDA for the prevention of miscarriage in 1947. It was prescribed to pregnant women with a history of spontaneous abortion (but was later discovered to be ineffective). In 1971, a report in the "New England Journal of Medicine" warned about a link between the use of DES during pregnancy in the 1950s and the development of a rare vaginal cancer in their daughters. It was later discovered that sons exposed to DES during pregnancy have an increased risk for genital anomalies like undescended testicles and epididymal cysts.[13]

In 1977, FDA issued guidelines that recommended the exclusion of women of childbearing potential from early phases of drug trials until safety and effectiveness were established, except in the case of life-threatening diseases. Public sentiment and litigation against drug manufacturers compounded the reluctance to include women of childbearing potential in drug studies. Women's health advocacy groups argued that the exclusion assumed "that women cannot take steps to avoid becoming pregnant," decided for women "that protecting the fetus outweighed other possible interests," and was unethical based on the principle of autonomy.[14] However, exclusion had become the norm.

The benefits of inclusion in research studies became more evident during the Acquired Immuno-Deficiency Syndrome (AIDS) crisis in the 1980s. Women began to question the accuracy of data derived from studies performed on men when applied to women. In addition, women with AIDS were angry that experimental drugs available to men in clinical studies were not available to them. They demanded a "right to choose for themselves whether to take on the health risks of drug research."[15] The realization grew that "policies designed to protect certain populations from research risk may actually expose these populations to a greater risk of another kind: a lack of data about their health."[16] Through activism and advocacy, regulations evolved until all women, even pregnant ones, were permitted to participate in clinical trials.

Since 1981, rules of ethics about biomedical and behavioral research have been codified in Health and Human Services Title 45 Code of Federal Regulations (CFR) Part 46 (Public Welfare) Subparts A, B, C, and D also known as "The Common Rule."[17] US policy regarding the inclusion of pregnant women in federally funded research is in Subpart B (see Table 1.1.).

Essentially, these regulations (with some caveats) state that a pregnant woman may participate in a research study if:

- studies on animals and nonpregnant women have provided data that help define the potential risk to the mother/baby,
- the research will benefit either the mother or the baby,
- the study method provides the least possible risk to the mother/baby,
- the mother gives consent, or
- the mother and the father give consent if the research benefits only the baby.

While the CFR *allows* for the participation of pregnant women in research studies, it does not *mandate* their inclusion. Thus, the *practice* of including pregnant women in research studies remains uncommon.[18]

Proponents of inclusion

In October of 2007, Anne Drapkin Lyerly, an obstetrician/gynecologist and ethicist, then at the Trent Center for Bioethics at Duke University, along with colleagues Margaret Little (Kennedy Institute of Ethics, Georgetown

TABLE 1.1 The common rule—subpart B.

TITLE 45—PUBLIC WELFARE PART 46_PROTECTION OF HUMAN SUBJECTS

Subpart B Additional Protections for Pregnant Women, Human Fetuses and Neonates Involved in Research

Sec. 46.204 Research involving pregnant women or fetuses

Pregnant women or fetuses may be involved in research if all of the following conditions are met:

(a) Where scientifically appropriate, preclinical studies, including studies on pregnant animals, and clinical studies, including studies on nonpregnant women, have been conducted and provide data for assessing potential risks to pregnant women and fetuses;

(b) The risk to the fetus is caused solely by interventions or procedures that hold out the prospect of direct benefit for the woman or the fetus; or, if there is no such prospect of benefit, the risk to the fetus is not greater than minimal and the purpose of the research is the development of important biomedical knowledge which cannot be obtained by any other means;

(c) Any risk is the least possible for achieving the objectives of the research;

(d) If the research holds out the prospect of direct benefit to the pregnant woman, the prospect of a direct benefit both to the pregnant woman and the fetus, or no prospect of benefit for the woman nor the fetus when risk to the fetus is not greater than minimal and the purpose of the research is the development of important biomedical knowledge that cannot be obtained by any other means, her consent is obtained in accord with the informed consent provisions of subpart A of this part;

(e) If the research holds out the prospect of direct benefit solely to the fetus then the consent of the pregnant woman and the father is obtained in accord with the informed consent provisions of subpart A of this part, except that the father's consent need not be obtained if he is unable to consent because of unavailability, incompetence, or temporary incapacity or the pregnancy resulted from rape or incest;

(f) Each individual providing consent under paragraph (d) or (e) of this section is fully informed regarding the reasonably foreseeable impact of the research on the fetus or neonate;

(g) For children as defined in Sec. 46.402(a) who are pregnant, assent and permission are obtained in accord with the provisions of subpart D of this part;

(h) No inducements, monetary or otherwise, will be offered to terminate a pregnancy;

(i) Individuals engaged in the research will have no part in any decisions as to the timing, method, or procedures used to terminate a pregnancy; and

(j) Individuals engaged in the research will have no part in determining the viability of a neonate.

University), Lisa Harris (University of Michigan), and Ruth Faden (Berman Institute of Bioethics, Johns Hopkins University) participated in a panel discussion at the American Society for Bioethics and Humanities annual meeting.[19] The session was entitled, "The Second Wave: A Moral Framework for Clinical Research with Pregnant Women." The panel identified the "lack of an adequate moral framework for guideline development" to be a significant barrier to the inclusion of pregnant women in research—and one that sustains the near-universal presumption of exclusion—to the detriment of maternal and fetal health.

The following year, Lyerly, Little, and Faden published a paper entitled, "The Second Wave: Toward Responsible Inclusion of Pregnant Women in Research."[20] The paper highlighted the reasons that pregnant women should be included in research: to gain knowledge about how to effectively treat pregnant women and keep the fetus safe, to prevent harm from withholding treatment that might be effective, for pregnant women to have the ability to decide for themselves whether to participate in studies, and to have access to the medicines available through research. The authors proposed that the exclusion of pregnant women from a research study should need to be justified—as opposed to the current practice of assuming exclusion unless inclusion can be justified. They proposed that the acceptance of inclusion and the justification for appropriate exclusion should be the norm and would benefit from an articulated ethical framework.

In April of 2009, the three authors, joined by Jason Umans from Georgetown University Hospital, sponsored an invitation-only workshop to discuss the costs of excluding and the barriers to including pregnant women in medical research. Various leaders in the field of women's health care research participated including those in academia (Georgetown, Johns Hopkins, Duke), government (National Institutes of Health [NIH], FDA), and additional ethicists, women's health advocates, and health care providers. The workshop sought to design actions to address priority issues and to discover what important information was still missing from the debate.

Two key pieces of missing information related to the pharmaceutical industry were identified by the workshop participants: (1) the perception of litigation risk (how influential is it and is it a real risk?) and (2) the role of the pharmaceutical industry (how influential is it in affecting the outcome of the debate?).[21]

It is worth noting that participation in the workshop was by invitation only—and that no one from the pharmaceutical industry had been invited.

The participation of pregnant women in medical research is a sensitive topic that could be easily misunderstood and misconstrued by external audiences, particularly those with specific agendas regarding fetal protection. Describing the negative impact of nonparticipation is complex and is not easily captured in sound bites. Workshop sponsors may have desired to limit the participants to those who understood the issues a priori and who could contribute to formulating actions that would further the agenda, including a plan to bring industry on board. However, the lack of an invitation could also be perceived as (1) a presumption of an adversarial position by industry and (2) a missed opportunity to acknowledge, educate, and involve a major stakeholder in the debate.

In the US, the pharmaceutical industry has replaced the NIH as the leading funder and conductor of clinical research studies.[22,23] There is a dearth of published information from the industry regarding pregnant women and drug studies. The extent of pregnant women's exclusion has not been quantified, nor has the industry's rationale for their exclusion been articulated.

In order to improve the treatment of medically compromised pregnancies, proponents (such as ethicists, women's health advocates, and health care providers in academia, government, and the health care system) want a change in practice towards a rational inclusion of pregnant women in clinical research. In order to accomplish this change, it would be prudent to know the current practices and attitudes of the pharmaceutical industry.

The Second Wave Consortium is not the only group of stakeholders who promote the need to perform research during pregnancy. Nor is the debate limited to the US. A commentary addressed to the OB/GYN community in the British Journal of Obstetrics and Gynecology in 2004 concluded that, "there needs to be a serious and ongoing debate about therapeutic research in the pregnant population and a consensus needs to be reached as to what levels of risk might be considered reasonable."[24]

Several documents from international agencies—whose raison d'etre is the promulgation of ethical standards for research activities—explicitly promote the inclusion of pregnant women in clinical research.[25]

In 2002, Guideline 17: Pregnant women as research participants of the CIOMS International Ethical Guidelines for Biomedical Research Involving Human Subjects[26] stated:

Pregnant women should be presumed to be eligible for participation in biomedical research. Investigators and ethical review committees should ensure that prospective subjects who are pregnant are adequately informed about the risks and benefits to themselves, their pregnancies, the foetus and their subsequent offspring, and to their fertility.

In 2017, Guideline 19: Pregnant and Breastfeeding Women as Research Participants of the CIOMS guidance[27] was revised to read:

A direct consequence of the routine exclusion of pregnant women from clinical trials is their use of medications (both prescription and non-prescription) lacking data from clinical trials about the potential individual benefits and harms to themselves, their fetuses and their future children. Therefore, after careful consideration of the best available relevant data, it is imperative to design research for pregnant and breastfeeding women to learn about the currently unknown risks and potential individual benefits to them, as well as to the fetus or nursing infant.

United Nations (UN)AIDS/World Health Organization (WHO) 2005 Guidance[28] Point 9 states:

Researchers and trial sponsors should include women in clinical trials in order to verify safety and efficacy from their standpoint, including immunogenicity in the case of vaccine trials, since women throughout the life span, including those who are sexually active and may become pregnant, be pregnant or be breastfeeding, should be recipients of future safe and effective biomedical HIV prevention interventions. During such research, women's autonomy should be respected and they should receive adequate information to make informed choices about risks to themselves, as well as to their foetus or breastfed infant, where applicable.

FDA guidance documents, while not binding, do provide standards and expectations and are usually closely adhered to by industry. Stated Kessler, "The FDA believes that it is critical to obtain a broad range of views on these matters from the public as well as from experts in the fields of medicine, health care, ethics, and the law, and we are committed to facilitating that exchange."[29] As with most new guidance documents, the release of the new FDA guidance in April 2018 was followed by a 60-day comment period where members of the public and other interested parties, including the pharmaceutical industry, could provide feedback to the agency on its contents. Following the comment period, the document is then re-reviewed and revised within FDA, and ultimately a final guidance document is published. The process can sometimes take several years to complete. Hopefully it will take less than the 25 years it took to draft.

Significance of the issue

Studies indicate that over 60% of pregnant women are prescribed one or more drugs (not including vitamins) during their pregnancies.[30,31] Inadvertent fetal exposure to acute or maintenance medication by women who do not yet realize that they are pregnant occurs frequently, as about half of all US pregnancies are unplanned.[32] But we have little data on the safety and efficacy of most medications when they are used during pregnancy. A recent study found that safety in pregnancy was unknown for over 80% of the 468 drugs marketed in a recent 20-year period, due to insufficient human data.[33] This leaves the clinician and the patient not knowing how to interpret the little data that do exist—whether to take a potentially effective medication or not; the effect it may have on the woman, the fetus, or the pregnancy; or whether or not to terminate a pregnancy based on the exposure. Nor is it always apparent what the negative consequences will be if she discontinues a medication, takes a lower dose of a medication, or takes a different medication.

In my experience conducting pregnancy registries in a pharmaceutical company, I have spoken to women who had been advised by their physicians to consider terminating a pregnancy during which they had inadvertently used a medication or received a vaccine. None of the medications involved were suspected of causing birth defects. The desire to decrease their liability risk if the infant was born with a birth defect is a potential motivation for such advice—but this is speculative. The number of women so advised and the number of pregnancy terminations resulting from inadvertent exposure to medication during pregnancy are unknown.

Experimental drugs with unknown teratogenic potential would rarely be tested on pregnant women in the first trimester. "For humans, the teratogenic period is relatively short, lasting from implantation of the embryo in the uterus, which occurs 5 to 7 days after conception, until the 8th week of human development..."[34] Unfortunately, however, this is the most common timing of inadvertent pregnancy exposures to marketed drugs—prior to the mother's suspicion of pregnancy at her first missed menstrual period.

Overall there is an approximate 3% risk of having a baby with a birth defect.[35] Most of the causes of these congenital anomalies are unknown and medication exposures are known to induce a very small percent.[36] In fact, the vast majority of drugs and vaccines do not cause fetal harm.[37] Preclinical animal testing has evolved greatly since the tragic impact of thalidomide and DES and, with one exception, all drugs that cause birth defects in humans have been shown to induce defects in animals as well.[38] However, it remains difficult to identify rarely-occurring defects that are caused by drugs. For example, it was only recently ascertained that ACE inhibitors, which have been on the market for over 20 years and were considered to be safe for use in the first trimester of pregnancy, increase the risk for cardiovascular and central nervous system defects.[39] So, while the risk is small for most exposures, the uncertainty remains. This situation leads to fear and potential overreaction by health care providers and pregnant women. Studies indicate that women and their health care providers tend to over-estimate the risk of medication-induced birth defects.[40]

A tragic example of the consequences of such fears is the story told to me by a health care provider whose 7-month-pregnant patient discontinued her asthma medications so as to avoid exposing her baby in utero. She subsequently experienced an acute asthma exacerbation and died. The asthma medications she was taking are recommended to be continued during pregnancy because their risk to the fetus is less than the risk of poorly controlled asthma, as this tragic story illustrates.

It is not only safety information that is lacking; there is a dearth of efficacy information as well. "Many physiological changes that women experience during pregnancy — such as increased plasma volume, body weight, body fat, metabolism and hormone levels — make it impossible to calculate dosage and efficacy information by extrapolating from data on men and nonpregnant women."[41] Only by conducting research on women in different trimesters of pregnancy can knowledge of dosing, timing, and efficacy be gained.

The potential impact of a lack of data for drug efficacy during pregnancy is illustrated by the 2002 recommendation by ACOG of the use of

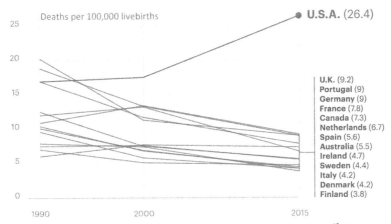

FIGURE 1.1   Maternal mortality in the US and other High Income Countries.[43] Source: *From "Global, regional, and national levels of maternal mortality, 1990–2015: a systematic analysis for the Global Burden of Disease Study 2015,"* The Lancet. *Only data for 1990, 2000 and 2015 was made available in the journal.*

amoxicillin by pregnant women for anthrax post-exposure prophylaxis. Subsequent study, the results of which were published in 2007, showed that this dosage and frequency recommendation was ineffective for pregnant and post-partum women and no studies are available for ciprofloxacin or doxycycline, the alternative antibiotics.[42]

Finally, the United States has been battling for years to bring down the rate of pregnancy-related deaths. The US rate of maternal death is higher than the rate in every other developed country in the world—by a large margin—and the rate is growing (see Fig. 1.1).

Increased morbidity, inadequate treatment, the unnecessary termination of wanted pregnancies maternal deaths—these are some of the consequences of the lack of information about the safety and efficacy of medications used to treat illness during pregnancy.

Ethical concerns

Concern about fetal safety is the primary motivation against researchers designing studies for pregnant women, against investigators including pregnant women, and against clinicians approaching pregnant women about participating in research studies—as well as the primary reason that pregnant women themselves decline to participate.

Clinical trials have traditionally excluded pregnant women from participation due to this concern. But the ethics of this exclusion are subject to challenge due to the consequential lack of information on the safety and efficacy of medications to treat the multitude of medical conditions that occur during pregnancy.

Which leaves an ethical conundrum—we want to improve the health of pregnant women and their babies, yet to do so we need to do research on pregnant women and their babies, which might harm pregnant women and their babies, and therefore we cannot do research to improve the health of pregnant women and their babies. Yet we do research with attendant risks on men and nonpregnant women. This begs the question, has too much emphasis on nonmaleficence in the pregnant population precluded us from achieving the health benefits of scientific research that have accrued to the nonpregnant population?

The dilemma arises from a conceptual evolution in modern (i.e., developed nations') obstetrical practice. With technological advances that allow health care providers to see, hear, and actually touch the fetus inside its mother's uterus, many have advanced the notion that the fetus is a patient in its own right. Prior to these individualizing capabilities, the mother-fetal dyad was widely considered to be a "patient package"—a unit. The fetus could only be evaluated indirectly—via palpation and measurement of the mother's abdomen, testing of the mother's blood, external fetal monitoring, etc. It was not until labor revealed the fetal scalp through the mother's dilating cervix that direct access to fetal blood was possible. The woman was considered to be somewhat of a barrier to be overcome in order to assess fetal status.[44] In that milieu, the dependency of the fetus on the mother and the mother as the primary patient was obvious. With the advent of ultrasound, amniocentesis, umbilical cord blood testing, fetal surgery, and perinatology, however, many have advanced the notion of the fetus as a patient in its own right—sometimes as an individual, conceptually separable patient. Technological developments have led directly to the concepts of fetal rights and fetal 'autonomy' in a clinical as well as a political sense.[45] Our ethical conceptions evolve and adapt to encompass advances in technology.

Nonmaleficence, the ethical dictate to avoid doing harm, may influence us to include or exclude pregnant women in studies. Beneficence, to do good, may influence us to include or exclude pregnant women in studies. But justice, the "equitable distribution of the burdens and benefits of research,"[46] requires that we find a way to obtain evidence-based knowledge to formulate best practices to treat pregnant women and their fetuses. (For further discussion on ethics, see Chapter 3, The ethics involved.)

Aims of the book

In doing the investigations that went into this book, I sought to add research and scholarship to the debate about the inclusion of pregnant women in clinical trials—one of the aims of the Second Wave Consortium's effort to change the status quo.

I sought to isolate, articulate, and communicate the opinions of selected representatives from the pharmaceutical industry and related organizations about the inclusion and exclusion of pregnant women in clinical research studies. By speaking to experts who work in, or interact with, the industry, I hoped to identify attitudes and practices regarding pregnant women in clinical research including opportunities for, and barriers to, broadening their safe inclusion.

Specifically, my aims were to:

- Quantify the frequency of the participation of pregnant women in current pharmaceutical company-based studies by accessing the inclusion and exclusion criteria listed on ClinicalTrials.gov. (The Food and Drug Modernization Act of 2007 mandates that all federally and privately funded clinical trials be posted on the NIH website, ClinicalTrials.gov).
- Raise the issue to selected pharmaceutical industry representatives and related organizations to heighten their awareness of the issue and the debate.
- Isolate the concerns of the pharmaceutical industry representatives about including pregnant women in clinical trials to further our understanding of the reasons for exclusion, and potential barriers to their inclusion.
- Isolate potential opportunities for inclusion of pregnant women in clinical trials from the pharmaceutical industry representatives' perspectives.
- Describe the pharmaceutical industry and related organizations' perceptions of litigation risk.

Upon release of the draft FDA guidance document in April 2018, the findings discussed in this book were submitted to FDA in response to its public call for comment. (All nonconfidential comments submitted in the Call for Comment are accessible to the public at https://www.regulations.gov by searching on the docket number (FDA-2018-D-1201) or by entering Pregnant Women in the search box and scrolling through the findings.)

The findings were also sent to PhRMA (Pharmaceutical Research and Manufacturers' Association), to provide them with additional information with which to respond to FDA's call for comment, and, based on the results of our interviews, potential solutions to the issue of including pregnant women in clinical studies moving forward.

The information was also shared with the stakeholders who participated in the interviews so that they could share the results within their respective companies and organizations. In that way, the information may contribute to debate in multiple settings and disciplines.

In addition, the results of the quantitative assessment were published in the peer-reviewed scientific literature (Obstetrics & Gynecology 2013; 122 (5):1077–1081). The paper provided a measure of the prevalence of the exclusion of pregnant women from clinical trials.[47] (These results are discussed in Chapter 4, A measure of exclusion.) It was hoped that this information would increase awareness and initiate discussion, in advance of the "pending" release of a guidance document from FDA. The publication contains a measure of pregnant women's current exclusion from clinical research which can now serve as a baseline against which to evaluate the impact of the new guidance on increasing their inclusion.

This book seeks to present these findings to a broader audience. As Dr. Kessler said back in 1993, "it is critical to obtain a broad range of views on these matters from the public as well as from experts in the fields of medicine, health care, ethics, and the law, and we are committed to facilitating that exchange."[48] By compiling the research results and investigative findings in this book, this author seeks to facilitate that exchange as well. I think that the general public, along with many of the "experts" Kessler cites, are currently under-informed about this important topic. Most people, understandably, give little thought to the need for better information on drugs that are used by ailing pregnant women—until they or their loved ones are ailing pregnant women themselves. I hope the book will raise awareness, challenge current thinking, and facilitate change towards a more responsible policy and practice that will ultimately improve the evidence-based treatment of pregnant women.

Notes

1. Zajicek, A., & Giacoia, G. P. (2007). Clinical pharmacology: Coming of age. *Clinical Pharmacology and Therapeutics, 81*(4), 481–482.
2. U.S. Department of Health and Human Services, Public Health Service, National Institutes of Health, Office of Research on Women's Health. (2011). *Enrolling pregnant women: Issues in clinical research.* Bethesda, MD: National Institutes of Health. Retrieved from <https://orwh.od.nih.gov/sites/orwh/files/docs/ORWH-EPW-Report-2010.pdf>.
3. Lyerly, A. D., Little, M. O., & Faden, R. R. (2011). Reframing the framework: Toward fair inclusion of pregnant women as participants in research. *The American Journal of Bioethics, 11*(5), 50–52. <https://doi.org/10.1080/15265161.2011.560353>.
4. Feibus, K., & Goldkind, S. F. (2011, May 17). *Pregnant women and clinical trials: Scientific, regulatory, and ethical considerations.* Oral presentation at the Pregnancy and Prescription Medication Use Symposium, Silver Springs, MD.
5. Mastroianni, A. C., Faden, R., & Federman, D. (Eds.). (1994). *Women and health research: Ethical and legal issues of including women in clinical studies.* Washington, DC: National Academy Press, Institute of Medicine.
6. *Council for International Organizations of Medical Sciences. CIOMS international ethical guidelines for health-related research involving humans.* (2016). Retrieved from <https://cioms.ch/wp-content/uploads/2017/01/WEB-CIOMS-EthicalGuidelines.pdf>.
7. American College of Obstetricians and Gynecologists. (2015). ACOG Committee Opinion No. 646: Ethical considerations for including women as research participants. *Obstetrics & Gynecology, 126*, e100–e107. <https://doi.org/10.1097/AOG.0000000000001150>.
8. Piper, J. M., Ray, W. A., & Rosa, F. W. (1992). Pregnancy outcome following exposure to angiotensin-converting enzyme inhibitors. *Obstetrics and Gynecology, 80*, 429–432.
9. Kessler, D. A., Merkatz, R. B., & Temple, R. (1993). Authors response to Caschetta, M. B., Chavkin, W., & McGovern, T. Correspondence: FDA policy on women in drug trials. *New England Journal of Medicine, 329*(24), 1815–1816. <https://doi.org/10.1056/NEJM199312093292414>.

10. U.S. Food and Drug Administration. (2018). Pregnant women: Scientific and ethical considerations for inclusion in clinical trials. Guidance for industry. Retrieved from: <https://www.fda.gov/ucm/groups/fdagov-public/@fdagov-drugs-gen/documents/document/ucm603873.pdf>.

11. Kim, J. H., & Scialli, A. R. (2011). Thalidomide: The tragedy of birth defects and the effective treatment of disease. *Toxicological Sciences, 122*(1), 1–6. <https://doi.org/10.1093/toxsci/kfr088>.

12. Rehman, W., Arfons, L. M., & Lazarus, H. M. (2011). The rise, fall and subsequent triumph of thalidomide: Lessons learned in drug development. *Therapeutic Advances in Hematology, 2*(5), 291–308. <https://doi.org/10.1177/2040620711413165>.

13. *National Cancer Institute. Diethylstilbestrol (DES) and cancer.* (2011). Retrieved from <https://www.cancer.gov/about-cancer/causes-prevention/risk/hormones/des-fact-sheet>.

14. *U.S. Food and Drug Administration. Gender Studies in product development: Historical overview.* (2018). Retrieved from <https://www.fda.gov/ScienceResearch/SpecialTopics/WomensHealthResearch/ucm134466.htm>.

15. Institute of Medicine (U.S.) Committee on Ethical and Legal Issues Relating to the Inclusion of Women in Clinical Studies, Mastroianni, A. C., Faden, R., & Federman, D. (Eds.). (1994). *Women and health research: Ethical and legal issues of including women in clinical studies: Vol. I* (p. 6). Washington, DC: National Academies Press (US), Legal Considerations. Retrieved from <https://www.ncbi.nlm.nih.gov/books/NBK236532/>.

16. *U.S. Food and Drug Administration. Gender studies in product development: Historical overview.* (2018). Retrieved from <https://www.fda.gov/ScienceResearch/SpecialTopics/WomensHealthResearch/ucm134466.htm>.

17. Code of Federal Regulations, Title 45, Part 46, Subpart B. (2016, October 1). *U.S. Government Printing Office via GPO Access* (pp. 140–143). Retrieved from <https://www.gpo.gov/fdsys/pkg/CFR-2016-title45-vol1/pdf/CFR-2016-title45-vol1-part46.pdf>.

18. Baylis, F. (2010). Opinion: Pregnant women deserve better. *Nature, 465,* 689–670.

19. Lyerly, A. D., Faden, R. R, Harris, L., & Little, M. O. (2007). *American Society for Bioethics and Humanities Annual Meeting. Panel session: The Second Wave: A moral framework for clinical research with pregnant women.* Retrieved from <http://asbh.confex.com/asbh/2007/techprogram/P6364.HTM>.

20. Lyerly, A. D., Little, M. O., Faden, R. (2008). The Second Wave: Toward responsible inclusion of pregnant women in research. *International Journal of Feminist Approaches in Bioethics, 1*(2), 5–22.

21. Lyerly, A. D. (2010, November 12). Personal communication.

22. Wendler, D. (2017, Spring). The ethics of clinical research. In E. N. Zalta, (Ed.). *The Stanford Encyclopedia of Philosophy.* Retrieved from <https://plato.stanford.edu/entries/clinical-research/#WhatClinRese>.

23. Chopra, S. S. (2003). Industry funding of clinical trials: benefit or bias? *Journal of the American Medical Association, 290*(1), 113–114. <https://doi.org/10.1001/jama.290.1.113>.

24. Lupton, M. G. F., & Williams, D. J. (2004). The ethics of research in pregnant women: Is maternal consent sufficient? *British Journal of Obstetrics and Gynecology, 111,* 1307–1312. <https://doi.org/10.1111/j.1471-0528.2004.00342.x>.

25. Macklin, R. (2010). The art of medicine: Enrolling pregnant women in biomedical research. *The Lancet, 375,* 632–633. <https://doi.org/10.1016/S0140-6736(10)60257-7>.

26. *Council for International Organizations of Medical Sciences. CIOMS international ethical guidelines for health-related research involving humans.* (2012). Retrieved from <https://cioms.ch/wpcontent/uploads/2016/08/International_Ethical_Guidelines_for_Biomedical_Research_Involving_Human_Subjects.pdf>.

27. *Council for International Organizations of Medical Sciences. CIOMS international ethical guidelines for health-related research involving humans.* (2016). Retrieved from <https://cioms.ch/wp-content/uploads/2017/01/WEB-CIOMS-EthicalGuidelines.pdf>.

28. UNAIDS/World Health Organization, Joint United Nations Programme on HIV/AIDS. (2007). *Ethical considerations for biomedical HIV prevention trials: Guidance document.* Geneva. Retrieved from <http://www.unaids.org/sites/default/files/media_asset/jc1399_ethical_considerations_en_0.pdf>.

29. Kessler, D. A., Merkatz, R. B., & Temple, R. (1993). Authors response to Caschetta, M. B., Chavkin, W., & McGovern, T. Correspondence: FDA policy on women in drug trials. *New England Journal of Medicine, 329*(24), 1815−1816. <https://doi.org/10.1056/NEJM199312093292414>.

30. Andrade, S. E., Gurwitz, J. H., & Davis, R. L. (2004). Prescription drug use in pregnancy. *American Journal of Obstetrics and Gynecology, 191*, 398−407. <https://doi.org/10.1016/j.ajog.2004.04.025>.

31. Glover, D. D., Amonkar, M., Rybeck, B. F. & Tracy, T. S. (2003). Prescription, over-the-counter, and herbal medicine use in a rural obstetric population. *American Journal of Obstetrics & Gynecology, 188*(4), 1039−1045.

32. Henshaw, S. K. (1998). Unintended pregnancy in the United States. *Family Planning Perspectives, 30*, 24−29. <https://doi.org/10.1001/jama.289.13.1697>.

33. Lo, W. Y., & Friedman, J. M. (2002). Teratogenicity of recently introduced medication in human pregnancy. *Obstetrics and Gynecology, 100*, 465−473. <https://doi.org/10.1016/S0029-7844(02)02122-1>.

34. Frederiksen, M. C. (2008). Commentary: A needed information source. *Clinical Pharmacology & Therapeutics, 83*(1), 22−23. <https://doi.org/10.1038/sj.clpt.6100438>.

35. Centers for Disease Control and Prevention. (2018). *Birth defects.* Washington, DC: Author. Retrieved from <http://www.cdc.gov/ncbddd/birthdefects/data.html>.

36. Brent, R. L. (2004). Environmental causes of human congenital malformations: The pediatrician's role in dealing with these complex clinical problems caused by a multiplicity of environmental and genetic factors. *Pediatrics, 113*(4), 957−968.

37. Koren, G., Pastuszak, A., & Ito, S. (1998). Drugs in pregnancy. *New England Journal of Medicine. 338*, 1128−1137. <https://doi.org/10.1056/NEJM199804163381607>.

38. Koren, G., Pastuszak, A., & Ito, S. (1998). Drugs in pregnancy. *New England Journal of Medicine. 338*, 1128−1137. <https://doi.org/10.1056/NEJM199804163381607>.

39. Cooper, W. O., Hernandez-Diaz, S., Arbogast, P. G., Dudley, J. A., Dyer, S., Gideon, et al. (2006). Major congenital malformations after first-trimester exposure to ACE inhibitors. *New England Journal of Medicine, 354*, 2443−2451. <https://doi.org/10.1056/NEJMoa055202>.

40. Koren, G., Bologa, M., Long, D., Feldman, Y., & Shear, N. H. (1989). Perception of teratogenic risk by pregnant women exposed to drugs and chemicals during the first trimester. *American Journal of Obstetrics & Gynecology, 160*, 1190−1194.

41. Baylis, F. (2010). Opinion: Pregnant women deserve better. *Nature, 465*, 689−670.

42. Andrew, M. A., Easterling, T. R., Carr, D. B., Shen, D., Buchanan, M., Rutherford, T, et al. (2007). Amoxicillin pharmacokinetics in pregnant women: Modeling and simulations of dosage strategies. *Clinical Pharmacology & Therapeutics, 81*(4), 547−556. <https://doi.org/10.1038/sj.clpt.6100136>.

43. Martin, N., & Montagne, R. (2017). U.S. has the worst rate of maternal deaths in the developed world. National Public Radio/WHYY. Retrieved from <https://www.npr.org/2017/05/12/528098789/u-s-has-the-worst-rate-of-maternal-deaths-in-the-developed-world> citing Global Burden of Disease 2015 Maternal Mortality Collaborators. (2016). Global regional, and national levels of maternal mortality, 1990−2015: a systematic analysis for the Global Burden of Disease Study 2015. *The Lancet, 388*(10053), 1775−1812. <https://doi.org/10.1016/S0140-6736(16)31470-2>.

44. Rhoden, N. K. (1987). Informed consent in obstetrics: Some special problems. *Western New England Law Review, 9*(9)(1)/6, 67−88. Retrieved from <https://digitalcommons.law.wne.edu/lawreview/vol9/iss1/6>.

45. Rhoden, N. K. (1987). Informed consent in obstetrics: Some special problems. *Western New England Law Review*, 9(9)(1)/6, 67–88. Retrieved from <https://digitalcommons.law.wne.edu/lawreview/vol9/iss1/6>.
46. Levine, R. J. (1997). *Ethics and regulation of clinical research*. New Haven, CT: Yale University Press, cited in Weijer, C., Dickens, B., & Meslin, E. M. (1997). Bioethics for clinicians: 10. Research ethics. *Canadian Medical Association Journal*, 156(8), 1153–1157.
47. Shields, K. E., Lyerly, A. D. (2013). Exclusion of pregnant women from industry-sponsored clinical trials. *Obstetrics & Gynecology*, 122(5), 1077–1081. <https://doi.org/10.1097/AOG.0b013e3182a9ca67>.
48. Kessler, D. A., Merkatz, R. B., & Temple, R. (1993). Authors response to Caschetta, M.B., Chavkin, W., & McGovern, T. Correspondence: FDA policy on women in drug trials. *New England Journal of Medicine*, 329(24), 1815–1816. <https://doi.org/10.1056/NEJM199312093292414>.

Chapter 2

The rationales for and against inclusion

The Committee on the Ethical and Legal Issues Relating to the Inclusion of Women in Clinical Studies was convened by the Institute of Medicine (IOM) in 1992 at the request of the National Institutes of Health (NIH) Office of Research on Women's Health. The committee was asked to investigate, report findings, and propose recommendations to improve the inclusion of women, women of childbearing potential, and pregnant women in clinical studies. Their 1994 report on women and health research made several recommendations[1] including that pregnant women be presumed to be eligible for participation in clinical studies and that the decision about whether to participate or not should be made by the woman, following the provision and discussion of risk and benefit information by the investigator.

Over two decades later, this recommendation has not been implemented. Yet the risk of a medication causing a birth defect was unknown for over 90% of drugs approved in the United States between 1980 and 2000.[2] And, in a study that interviewed over 30,000 pregnant women, researchers found that between 1976 and 2008, the number of women taking four or more medications during pregnancy tripled and first-trimester use of medication increased by over 60%.[3]

I confirmed the lack of research involving pregnant women when, in a broad review of the medical literature, I found very few papers reporting the conduct or results of drug safety or efficacy studies that enrolled pregnant women. The studies I did find were mostly on lifestyle topics like diet, obesity, and exercise, on alternative therapies like moxibustion and acupuncture, or on behavioral factors like tobacco and substance abuse. I did locate studies that enrolled pregnant women with human immunodeficiency virus (HIV) and other sexually transmitted diseases and a few looking at cancer care and mental health treatment during pregnancy. But there were none that addressed common conditions like autoimmune disease, hypertension, or infectious diseases that could affect pregnant women and for which pharmacotherapy would be indicated.

Nor did I uncover any papers in the medical literature that described the evaluation of a subpopulation of pregnant women in a study that included both pregnant and nonpregnant subjects. Pregnant women were excluded

Pregnancy and the Pharmaceutical Industry. DOI: https://doi.org/10.1016/B978-0-12-818550-6.00002-7

from general clinical studies and studies specifically for and about pregnant women were rarely being conducted.

I found a number of papers about the practical and ethical problems associated with pregnant women's exclusion and addressing their need for inclusion; but I found no papers defending their exclusion. I speculate that, because exclusion is the status quo, there is little perceived need to discuss, rationalize, explain, or defend it. Only those interested parties who believed that a change is necessary felt the need to provide their rationale and to build support to alter the practice of exclusion.

Following all this searching, I felt that I needed a better understanding of the practice of, and the rationales for, excluding pregnant women from clinical studies. Since no papers were found that quantified the exclusion of pregnant women from research studies, the assumption that the practice is extensive needed to be verified. Many of the rationales I found for excluding pregnant women from clinical trials were extracted from papers on the subject of why we need to include them. Therefore, these findings may not reflect the actual thinking of the proponents of exclusion. My view was that I needed to directly ask those responsible for the exclusion [such as pharmaceutical companies and institutional review board (IRB) members] for their reasoning to verify the rationales stated by proponents of inclusion and to make the justifications explicit. With the release of the new US Food and Drug Administration (FDA) draft guidance for industry on the inclusion of pregnant women in clinical trials, it is even more imperative to explore the pharmaceutical industry's practices and perspectives on the issue.

The reason for not including pregnant women in clinical trials is often stated as, "Of course, we cannot ethically test drugs on pregnant women."[4] Yet there are robust ethical principles that support the arguments both for and against the participation of pregnant women in clinical research. Similar rationales were sometimes cited to support both inclusion and exclusion (e.g., fetal safety, legal risk). Because this topic is laden with ethical issues, I have categorized each identified rationale for and against inclusion by the ethical principle that best applies to the reasoning therein (see Tables 2.1 and 2.2).

TABLE 2.1 Rationales against the inclusion of pregnant women in clinical trials.

1 The uncertain effect of new drugs on the mother and/or the fetus
2 Litigation risk—because birth defects are relatively common, they may occur unrelated to the experimental drug exposure and result in spurious litigation
3 The number of pregnant women needed to participate in the study in order to achieve statistical significance is unachievable
4 Safer study designs are available
5 Alternative treatments are often available
6 Little return on investment
7 Regulations do not require inclusion

TABLE 2.2 Rationales for the inclusion of pregnant women in clinical research.

1 To acquire knowledge that improves the medical treatment of pregnant women and their offspring
2 To improve birth outcomes
3 To improve pregnant women's access to the benefits of clinical research
4 To improve the ethical acquisition of information about exposed pregnancies
5 Because regulations do not require the exclusion of pregnant women
6 Excluding pregnant women from participating in medical research is unethical and illegal—and may increase litigation risk
7 To follow the advice of experts in the field of women's health, law, and ethics

---

For further discussion on the ethical issues involved regarding pregnant women in clinical research see Chapter 3, The ethics involved.

Rationales against the inclusion of pregnant women in clinical research

"Primum non nocere—First, do no harm"[5]

Rationale 1: the uncertain effect of new drugs on the mother and/or the fetus[6]

Ethical rationale: nonmaleficence

The risk of unforeseen adverse effects on the woman, on the pregnancy, and on the fetus from exposure to an experimental compound is too uncertain to include pregnant women in clinical trials. This risk is one of the most frequently cited reasons for the exclusion of pregnant women from clinical research.[7,8,9,10,11]

The FDA acknowledges this issue. The "potential risks of fetal injury, the definition of circumstances under which such risks are justified, and the design of trials that will properly address the risks raise many challenging medical, scientific, legal, and ethical questions," stated David Kessler, former FDA Commissioner, in a 1993 editorial response.[12]

Assurance of drug safety is dependent on the size and composition of the population studied in the clinical trials. The size of the studies is dependent upon a number of factors including the burden of the disease in the population which affects the number of subjects available to participate. Other considerations include cost and urgency—some studies of diseases for which there are few or marginally effective treatments may require smaller sample sizes in order to get the product to patients more expeditiously. Even studies that are very large, enrolling thousands of subjects, cannot ensure the safety or efficacy of a drug when it is used by a pregnant woman unless many pregnant women were included in the trial.

Prior to allowing drug (or vaccine) testing on humans, preclinical studies are performed on animals to evaluate both efficacy and safety. Unfortunately, the results of animal studies do not always accurately predict the effects of treatment on human pregnancies.[13,14] While animal reproductive studies are very important in identifying potential teratogenic effects in human gestation, they are not definitive. A teratogen, from the Greek "teras" meaning monster, is any substance that may cause birth defects via a toxic effect on an embryo or fetus.[15] Positive findings of teratogenicity in animals do not mean the drug will cause birth defects in humans, and conversely, the absence of teratogenic effects in animals does not ensure safety for human fetuses.[16]

The number of unintended human pregnancies that occur in Phases II and III clinical trials are usually too few to provide definitive data. Often, pregnant subjects are dis-enrolled from the study upon confirmation of the pregnancy though pregnancy should be followed by the clinical staff until its outcome. Therefore, we cannot rely on animal testing and preliminary clinical trials to know, or to reassure pregnant women, that it would be safe for them to participate in clinical research.

Rationale 2: litigation risk—because birth defects are not uncommon, they may occur unrelated to experimental drug exposure and result in spurious litigation

Ethical rationale: financial stewardship

An increased risk of both warranted and spurious lawsuits against pharmaceutical companies and researchers is a commonly cited reason for the exclusion of pregnant women from clinical studies.[17,18,19,20] Professor Vanessa Merton calls this "tort phobia."[21]

According to population-based research studies, the risk of having a baby with a birth defect is about 3% for major congenital anomalies[22] (structural defects with surgical, medical, or serious cosmetic consequences) and up to 15% for minor anomalies (structural defects that are usually of no surgical, medical, or serious cosmetic consequence).[23] The risk of having a baby with a specific birth defect varies widely—from 1 in 100 infants for heart defects to 1 in 15,000−40,000 for rare disorders like achondroplasia.[24]

Therefore, if 100 pregnant women were to participate in a clinical study, it is likely that 2−4 babies in the study would be born with a major birth defect unrelated to their exposure to the experimental drug. However, the mother, the researcher, reviewers of the study findings, and litigators could erroneously conclude that the defect was a result of the exposure. The drug could be incorrectly labeled as teratogenic and the drug's manufacturer could be subject to litigation. This misinterpretation could occur even though teratogenic agents are understood to produce specific phenotypic effects (observable physical traits) depending on the time in gestation and dosage of the exposure. It would be unlikely for a drug to cause a cleft lip in one child and a club foot in another. The injury is usually a characteristic effect or cluster of effects that is identifiable and reproduceable, that is, other babies would be born with the same or similar malformations. But lawyers and juries do not always recognize the principles and complexities of teratogenesis.

There is an actual and unfortunate example of this phenomenon. Bendectin, a combination of pyridoxine (vitamin B6) and doxylamine (an antihistamine), which are both available without a prescription, was approved and had been shown to be effective for the treatment of nausea and vomiting of pregnancy (NVP). Despite having been extensively studied in animal, clinical, and epidemiologic studies[25] with no findings of measureable risk to the developing fetus,[26] the product was withdrawn from the market by the manufacturer in 1983 due solely to the burdens of litigation.[27] The product remained on the market in the United Kingdom and Canada where it is widely used by pregnant women for NVP. After a 10-year absence, the FDA-approved drug returned to the market in 2013 under a new trademark.

In addition to the absence of an effective treatment for NVP on the US market, an alleged consequence of this "litigation effect" is the reluctance of US pharmaceutical companies to develop drugs for use during pregnancy. The fact that only two medications—oxytocin and dinoprostone—have been approved for use in pregnancy between 1962 and 2010 supports the observation that US pharmaceutical companies are reluctant to develop drugs for the litigious obstetrical market.[28]

Rationale 3: the number of pregnant women needed to participate in the study in order to show efficacy may be unachievable

Ethical rationale: nonmaleficence

Clinical trials are conducted for two primary purposes—to measure efficacy and to evaluate safety. In order to efficiently observe the clinical endpoints that confirm or refute efficacy, the characteristics of subjects permitted to enroll are usually narrowly defined.

People with renal impairment, children, and the elderly, for example, are populations that are regularly excluded from initial trials in order to avoid added complexity. Sometimes additional trials for these specific subpopulations are conducted after initial studies on the more homogeneous population have established efficacy. In addition to the safety concerns, the complex physiologic changes associated with the advancing stages of gestation can be used to justify the exclusion of pregnant women from participation in early clinical trials. Blood volume, renal clearance, body mass index, and hormone levels fluctuate throughout the pregnancy making study results more complicated to evaluate and substantiate.

A trial specifically designed to evaluate efficacy in the pregnant population would be more likely to achieve results that advance evidence-based care than studies evaluating a small number of inadvertent pregnancies, but difficulty with the recruitment and retention of pregnant women may be an obstacle.[29] Baylis cautions that, "persuading pregnant women to take part in research can be difficult."[30]

Evaluating safety—specifically, measuring the potential for a new drug product to induce birth defects—would be even more difficult. "Populations of several thousand would be needed to assess if the background risk produced by a particular treatment changes the rate of birth defects in general. If we are interested in a specific birth defect that occurs at a rate of 1 in 1000 or fewer, then to demonstrate that a drug does not produce that specific birth defect would require treated and nontreated populations on the order of tens to hundreds of thousands of pregnant women."[31] Clearly these numbers are not feasible. Therefore, the evaluation of human teratogenicity cannot rely solely upon clinical trials but requires accumulated data from other sources.

Rationale 4: safer study designs are available

Ethical rationale: beneficence

One of the requirements of The Common Rule that permits pregnant women to participate in research is that the study be designed to provide the least possible risk to the mother and the fetus. Alternative methods of identifying drug-induced birth defects, including pregnancy registries, case studies, pharmacovigilance, and case–control studies,[32] though not definitive, do not subject pregnant women to the risks of clinical trials, including the risk of receiving a placebo instead of a potentially effective medication. These alternative study approaches are performed after the drug has been approved and is on the market. Greenwood and others suggest that epidemiological studies may be the only way to collect data about rare congenital anomalies and the long-term effects of a new drug entity.[33,34] These can only be done after the drug is approved and marketed. Once basic science, animal testing, and clinical study data have been collected and analyzed and have met FDA standards of benefit/risk analysis acceptable for license approval, then studies of the safety of use in human pregnancy can be performed on marketed drugs. It is safer for the mother and fetus if the safety of a product is as well-established as possible before it is used during pregnancy.

Rationale 5: alternative treatments are often available

Ethical rationale: beneficence

In fact, we can *never* assure pregnant women that it is safe to participate in clinical research, because we can never prove that a drug is safe in all women at all times. So it is prudent to err on the side of caution and not subject pregnant women and their fetuses to the risks of medical research. Instead, health care providers should continue to use best practice guidelines and the medical literature to prescribe treatment that has been shown to be safe and effective over time. Brent advises that the "obstetrician can avoid product liability litigation by not prescribing drugs that have reproductive risks for the mother or developmental risks for the developing embryo or fetus."[35] Only in situations where the current treatment is not effective and significant morbidity or mortality to the mother or the fetus is likely should we resort to the use of an experimental medication. If the condition is not life-threatening or medically significant, we can utilize palliative measures or encourage tolerance of short-term discomfort to ensure that we provide the safest prenatal environment possible to protect the fetus from iatrogenic harm.

Rationale 6: little return on investment

Ethical rationale: financial stewardship

Pharmaceutical companies are publically held entities that have a responsibility to shareholders to increase profit and decrease loss. We have discussed the difficulties associated with conducting clinical trials on pregnant women. Recruiting, enrolling, and retaining a sufficient number of pregnant women to ensure that their participation will be statistically significant and generalizable to a larger pregnant population would be costly and has little hope of success. Even if the benefit of the product can be shown to be greater than the risk of adverse effects, the market of pregnant women is relatively minimal. Therefore, the company may rationally decide that the actual cost of drug development and the potential cost of litigation exceed any potential financial gains.[36] The experience with Bendectin reinforces this case. Enrolling pregnant women would expose the pregnant participants and their nonconsenting fetuses to medical risk while also exposing the company to significant legal and financial risk.

Alternative study designs, such as those using marketed drugs, has its own dangers. "A company that performs studies on one of its already-approved drugs risks 'generating results that could destroy the value of the product rather than enhance it'."[37] This author acknowledges, however, that "finding that a drug is unsafe for use during pregnancy could leave its broader market unaffected."[38] But negative publicity generated by the correct or erroneous finding that a drug causes birth defects could affect sales particularly among women of childbearing potential.

Also, limiting enrollment in research studies to nonpregnant women is less complicated, less costly,[39] and more efficient. The long-term result is that "therapies will become available sooner and cost less."[40]

Rationale 7: regulations do not require inclusion

Ethical rationale: financial stewardship

The US regulations in the Common Rule[41] state that pregnant women may participate in research studies under certain conditions. They do not state that pregnant women must be included in research studies or that pregnant women must be given the option to participate in research studies. The decision as to whether to include or exclude pregnant women from studies is left to the sponsor of the study or the IRB that approves the study. Until regulators make the inclusion of pregnant women mandatory, sponsors will continue to avoid the potential legal and financial risks by mandating exclusion.[42,43]

Rationales for the inclusion of pregnant women in clinical trials

"Pregnant women get sick and sick women get pregnant"[44]

Rationale 1: to acquire knowledge that improves the medical treatment of pregnant women and their offspring

Ethical rationale: beneficence, nonmaleficence

Many authors agree that the primary reason to consider the inclusion of pregnant women in research studies is to provide evidence-based treatment guidelines to improve the health of pregnant women and their babies.[45,46,47,48] The lack of information about how to treat the more than 9 million pregnant women with chronic conditions and the millions more who develop new medical conditions during pregnancy is a significant problem.[49] Because they have not been systematically evaluated in pregnant women, practically all medications used to treat illness during pregnancy are prescribed without FDA approval—essentially off-label use.[50]

Obstetricians are forced to make medical decisions with their patients without practical information about drug efficacy and safety. Women experience many physiological changes during pregnancy including increases in plasma volume, body weight, and body fat, and changes in metabolism and hormone levels. Extrapolating effective dosages and fetal risks from data on men and nonpregnant women is impossible.[51]

Effective medical care is based upon trial and error—and the systematic collection and analysis of data from research conducted in vitro and in vivo over time. Dr. Lyerly, an obstetrician and bioethicist, argues that the whole "purpose of the enterprise of clinical research is to take responsible, limited, and calculated risks in order to garner evidence...."[52] The results of these efforts inform and guide clinical practice. Excluding pregnant women from participation in research studies on new medication results in a lack of knowledge about the effectiveness, the appropriate dosage, and the potential side effects of medication when used during pregnancy—a time when the patient is most concerned about safe and effective treatment.

This lack of knowledge can and does result in a number of adverse consequences for the medically compromised pregnancy, including withholding treatment, under-treatment, or overexposure of pregnant women and their fetuses.[53] Pregnant women are prescribed medications with "no real basis for predicting their effects."[54] Health care providers may be reluctant to prescribe, and pregnant women themselves may discontinue medications—both of which may lead to the lack of effective management of medical conditions during pregnancy. They may lower the dose of the medication thinking that it will decrease the exposure to the fetus. However, this can result in the exposure of a fetus with no therapeutic benefit to the mother. Conversely, standard dosages of some medications may result in overdosing of the pregnant woman due to physiologic changes during gestation.

Sadly, lack of knowledge can and does lead to the elective termination of wanted pregnancies based on an unwarranted fear of birth defects following the exposure. I have been told first-hand, and Kass et al. agree, that physicians have encouraged women to terminate pregnancies and pregnant women have terminated otherwise wanted pregnancies based on an inflated perception of the risk of teratogenicity—"despite the fact that fewer than 30 drugs are proven human teratogens"[55] and the percentage of birth defects caused by medication is very low.[56]

Ironically, since women are excluded from much research in an effort to protect the fetus from harm, "significant harm to the child may result from not providing [maternal] treatments. The number of cases in which medications are given inappropriately during pregnancy constitutes a fraction of the number in which indicated therapy is inappropriately withheld."[57] States Lott, "the benefits of barring pregnant women from participating in research may, in the end, harm expecting mothers and their foetuses more than their inclusion in clinical trials."[58]

Rationale 2: to improve birth outcomes

Ethical rationale: beneficence, nonmaleficence

According to CDC, the United States ranked 26th among 26 industrialized nations in infant mortality,[59] and 56th out of 225 countries worldwide.[60] The US infant mortality rate, the rate at which babies die before their first birthday, was 5.9 deaths per 1000 live births in 2016.[61] Contributing factors include disparities among racial and ethnic groups, congenital malformations, prematurity and low birth weight, and sudden infant death syndrome.[62] Medical research would help to improve the prevention and treatment of these and other life-threatening conditions.

Healthy babies are dependent upon healthy mothers and healthy pregnancies. Fetal health can be compromised by conditions that affect women in general (e.g., lupus) or conditions specific to pregnancy (e.g., preeclampsia) or conditions of fetal origin (e.g., Rh incompatibility). Lack of knowledge about the efficacy or negative impact of various medications constrains treatment options and restricts the abilities of health care providers to provide the best care possible. Birth outcomes are compromised. Therefore, improved fetal safety—often cited as a reason for the exclusion of pregnant women from research—can be just as effectively cited as a justification for the inclusion of pregnant women in research.[63] "Due to the underrepresentation of pregnant women in research, clinicians and women face treatment decisions in the context of a dearth of evidence about how drugs work in pregnant bodies, what doses are safe and effective ... and which drugs pose teratogenic risk for fetuses—a dearth that often leads to reticence to prescribe or take indicated drugs, to the detriment of maternal and fetal health."[64]

Rationale 3: to improve pregnant women's access to the benefits of clinical research

Ethical rationale: justice

"Restriction of trials to non-pregnant individuals excludes a class of potential beneficiaries and places them at an unfair disadvantage" state Lyerly et al.[65] Participating in a clinical trial can provide benefits such as possible therapeutic advantage, better outcome of disease, closer monitoring than in routine practice, getting attention for other ailments, better physical and laboratory health checks, superior physicians, labs, and testing, more contact with the providers, access to contacts for future health information, remuneration, and contributions to society.[66]

To Lupton and Williams, "pregnant women are often treated with drugs that have been superseded in every other branch of medicine ... because newer drugs have not been fully investigated in the pregnant population."[67] Clinical trials provide access to current potential advances in medicine and health care practice. Advocates for populations that have been excluded from participation in research studies (i.e., women, people living with HIV/ acquired immunodeficiency syndrome, and children) have fought for, and succeeded in achieving, inclusion. "The former complete exclusion of fertile women led to more deaths of women with HIV than men and eventual revision of exclusionist policies."[68,69] FDA has since revised its restrictions and now believes it is essential to include pregnant women when it is their only way to access potentially life-saving treatments that are under investigation.[70] Including pregnant women in clinical studies would improve their access to new medications and better health care that could improve their health and the health of the pregnancy.

Rationale 4: to improve the ethical acquisition of information about exposed pregnancies

Ethical rationale: nonmaleficence, autonomy

Ruth Macklin, one of the founders of the field of bioethics, states that the most compelling reason for the inclusion of pregnant women in clinical research "is the need for evidence gathered under rigorous scientific conditions, in which fewer women and their fetuses would be placed at risk than the much larger number who are exposed to medication once they come to the market."[71]

For the vast majority of pregnancies in which medications have been prescribed, the birth outcomes are never recorded. According to Berlin, "pregnant women who must take certain medications are essentially participating in an uncontrolled and unmonitored experiment for which the data will most likely never be assessed."[72] Hall adds the additional point that "the quality of informed consent is better in a research setting than in the post-marketing environment where prescriptions are written with little instruction and little follow-up is done."[73] "[T]he assumption seems to be that the researcher, or perhaps the IRB member, or perhaps a federal bureaucrat is the best choice to judge the net harm and benefit, risk and advantage, that would result from a pregnant woman's participation in a protocol."[74] Rather, the well-informed pregnant woman, who, by being pregnant, has not lost her ability to evaluate information, judge risk, or make decisions for herself and her fetus, should be the one who decides.[75,76]

Lack of knowledge about medication use during pregnancy has led to efforts to collect information about pregnancy outcomes from women who take medications in the postmarketing environment—after the products have been approved. For example, pregnancy registries, studies that evaluate birth outcomes from women who have used approved medications during their pregnancies, have been established for some medical conditions and for some medications. FDA has improved its insistence upon the establishment of a registry for any new drug that is anticipated to be used in the treatment of diseases that can affect pregnant women. But, warns Macklin, "surveillance activities ... lack the rigor of the scientific gold standard: a prospective, randomized clinical trial in which pregnant women are enrolled."[77] Pregnancy registries run by the pharmaceutical industry have variable enrollment criteria, use passive surveillance techniques, and take years to accumulate enough pregnancy exposures to identify safety signals or risks and do not address efficacy.

Rationale 5: regulations do not require the exclusion of pregnant women

Ethical rationale: justice

The Declaration of Helsinki states, "Populations that are underrepresented in medical research should be provided appropriate access to participation in research."[78] The US Common Rule "does little to promote research inclusion for pregnant women,"[79] perhaps the new guidance will do better.

According to Hall, "there is no regulatory reason for excluding pregnant women from many studies."[80] Historically, women of childbearing potential were excluded from participating in studies based on the study sponsor's over-interpretation of the regulations that had only excluded them from the first and earliest part of the second phase of studies.[81] The exclusion of pregnant women from participation may be, in part, based on a similar misinterpretation.

According to current regulation, pregnant women may be included in studies under certain circumstances. Various study designs have been proposed that can decrease the risk to the fetus while still providing for the inclusion of pregnant women. In fact, the IOM, in its 2010 report, "Women's Health Research: Progress, Pitfalls, and Promise,"[82] recommends that pregnant women be included unless there is a specific reason to exclude them.

Rationale 6: excluding pregnant women from participating in medical research is unethical and illegal—and may increase litigation risk

Ethical rationale: justice

Ethical conduct requires the inclusion of pregnant women in clinical trials, should they choose to participate. Lyerly (2011), citing Mastroianni et al., states that, "access to research, not just protection from its risks, is a constitutive part of the ethical mandates governing clinical research."[83] "Issues of justice," they continue, "are perhaps the most pressing."[84] "Women have the right—the same right as men—to decide for themselves (and, therefore, implicitly, for their potential offspring), whether it is prudent and morally right for them to participate in a given protocol, and women do not lose that right when they become pregnant," agrees Merton.[85]

McCullough et al. question the ethics of treating pregnant women in the absence of clinical study data. They state, "Until the risks and benefits of the different treatment options are quantified and weighed against each other, the continued use of ... drugs in these women without a sound evidence-base raises major clinical and ethical concerns."[86]

Writing in IRB: A Review of Human Subjects Research, Jacquelyn Kay Hall concluded that "excluding women from publicly paid benefits on the basis of their sex is illegal."[87] "There is no regulatory reason for excluding pregnant women from many studies" she writes, therefore "to exclude all pregnant women from the potential benefits of some protocols is illegal."[88] Vanessa Merton concurs, "[R]esearch sponsors in fact have more to fear in the way of potential liability from the exclusion of ... pregnant women ... than from their inclusion."[89]

Some view "the automatic exclusion of pregnant subjects as possibly more related to protecting the institution and investigator (from liability) than the subject or her unborn fetus (from possible harm),"[90] An alternative view is that it is the inadequate testing of a drug prior to marketing that increases a company's risk of liability for adverse effects. While there are few reported cases of damages awarded due to injury from inclusion in research, there are a number of cases where damages were awarded for claims of inadequate testing.[91]

One can imagine the liability claims for thalidomide were it released onto the market today with inadequate evaluation of its teratogenic potential. Had animal testing been performed to today's standards, the teratogenic potential of the drug would likely have been identified. It is interesting to consider that if pregnant women had been included in the clinical trials for thalidomide, as tragic as the initial cases of birth defects would have been, thousands of cases of the severe limb defects that occurred in exposed children would have been prevented worldwide.[92] Macklin says this "is a simple utilitarian calculation, an appropriate method for decision-making when the intention is to decrease the number of individuals exposed to potential harm."[93]

Rationale 7: to follow the advice of experts in the field of women's health, law, and ethics

Ethical rationale: justice, nonmaleficence

The American College of Obstetricians and Gynecologists Committee on Ethics,[94] the IOM,[95] the International Ethical Guidelines for Biomedical Research Involving Human Subjects of Council for International Organizations of Medical Sciences,[96] and the FDA[97] have all concluded that pregnant women can be appropriately included in clinical research. Respect for the autonomy of patients, beneficence, and justice in the selection of participants are three oft-cited ethical justifications for the inclusion of pregnant women in clinical studies.[98,99] Denying pregnant women the opportunity to enroll in research studies denies them the potential benefits of participation (improved treatment, enhanced medical care) and the opportunity to act altruistically and help other pregnant women.[100] "Increasingly, research ethics committees are encouraging researchers not to exclude this group of participants from research so long as appropriate safeguards are in place."[101]

In April 2009, subject matter experts in clinical practice, biomedical ethics, NIH, FDA, and others participated in a workshop (the Second Wave Consortium) on the topic of the inclusion of pregnant women in medical research. Their deliberations concluded, in part, with the following statement:

> *We believe that the current paucity of research on effective and safe treatment of pregnant women's illnesses is unethical. It is unfair and irresponsible to continue a system that compels physicians to use therapeutic agents in an uncontrolled experimental situation virtually every time they prescribe for pregnant women, and for women and the fetuses they carry to shoulder those risks whenever pregnancy is complicated by illness. As we learned in pediatric and geriatric research, if a population is going to use a medication it must be studied in that population. Pregnant women and the children they bear are best protected through responsible inclusion in research, not broad-based exclusion from it.[102]*

Notes

1. Institute of Medicine, Mastroianni, A. C., Faden, R., & Federman, D. (Eds.). (1994). *Women and health research: Ethical and legal issues of including women in clinical studies.* Washington, DC: National Academy Press.
2. Lo, W. Y., & Friedman, J. M. (2002). Teratogenicity of recently introduced medication in human pregnancy. *Obstetrics and Gynecology, 100,* 465–473. https://doi.org/10.1016/S0029-7844(02)02122-1.
3. Mitchell, A. A., Gilboa, S. M., Werler, M. M., Kelley, K. E., Louik, C., & Hernandez-Diaz, S. (2011). Medication use during pregnancy, with particular focus on prescription drugs: 1976–2008. *American Journal of Obstetrics & Gynecology, 205*(1), 51.e1–51.e8. https://doi.org/10.1016/j.ajog.2011.02.029.
4. Greenwood, K. (2010). The mysteries of pregnancy: The role of law in solving the problem of unknown but knowable maternal-fetal medication risk. *University of Cincinnati Law Review, 79,* 267–322.
5. Gillon R. (1985). Primum non nocere and the principle of non-maleficence. *British Medical Journal, 291,* 130. https://doi.org/10.1136/bmj.291.6488.130.
6. Mohanna, K., & Tunna, K. (1999). Withholding consent to participate in clinical trials: Decisions of pregnant women. *British Journal of Obstetrics and Gynaecology, 106,* 892–897. https://doi.org/10.1111/j.1471-0528.1999.tb08426.x.
7. Lyerly, A. D., Little, M. O., & Faden, R. (2008). The second wave: Toward responsible inclusion of pregnant women in research. *International Journal of Feminist Approaches in Bioethics, 1*(2), 5–22. https://doi.org/10.3138/ijfab.1.2.5.
8. American College of Obstetricians and Gynecologists. (2015). ACOG Committee Opinion No. 646: Ethical considerations for including women as research participants. *Obstetrics & Gynecology, 126,* e100–e107. https://doi.org/10.1097/AOG.0000000000001150.
9. Beran, R. G. (2006). The ethics of excluding women who become pregnant while participating in clinical trials of anti-epileptic medications. *Seizure, 15,* 563–570. https://doi.org/10.1016/j.seizure.2006.08.008.
10. Weijer, C. (1999). Selecting subjects for participation in clinical research: One sphere of justice. *Journal of Medical Ethics, 25,* 31–36.

11. Zajicek A., & Giacoia, G. P. (2007). Clinical pharmacology: Coming of age. *Clinical Pharmacology and Therapeutics, 81*(4), 481−482.

12. Kessler, D. A., Merkatz, R. B., & Temple, R. (1993). Author's response to Caschetta MB et al. correspondence: FDA policy on women in drug trials. *New England Journal of Medicine, 329*(24), 1815−1816. https://doi.org/10.1056/NEJM199312093292414.

13. Macklin, R. (2010). The art of medicine: Enrolling pregnant women in biomedical research. *The Lancet, 375,* 632−633. https://doi.org/10.1016/S0140-6736(10)60257-7.

14. Brent, R. L. (2004). Utilization of animal studies to determine the effects and human risks of environmental toxicants. *Pediatrics, 113,* 984−995.

15. Wikipedia. (2018). *Teratology.* <https://en.wikipedia.org/wiki/Teratology/> Retrieved 10.10.18.

16. Brent, R. L. (2004). Utilization of animal studies to determine the effects and human risks of environmental toxicants. *Pediatrics, 113,* 984−995.

17. Lyerly, A., Little, M. O., & Faden, R. (2012). Perspective: Pregnancy and clinical research. *The Hastings Center Report,38*(6), 53. https://doi.org/10.1353/hcr.0.0089.

18. Greenwood, K. (2010). The mysteries of pregnancy: The role of law in solving the problem of unknown but knowable maternal−fetal medication risk. *University of Cincinnati Law Review, 79,* 267−322.

19. Charo, R. A. (1993). Protecting us to death: Women, pregnancy, and clinical research trials. *Saint Louis University Law Journal, 38,* 135−187.

20. Kaposy, C., & Baylis, F. (2011). The Common Rule, pregnant women, and research: No need to "rescue" that which should be revised. *The American Journal of Bioethics, 11*(5), 60−62. https://doi.org/10.1080/15265161.2011.560360.

21. Merton, V. (1993). The exclusion of pregnant, pregnable, and once-pregnable people (a.k.a. women) from biomedical research. *American Journal of Law & Medicine, 19*(4), 369−451.

22. Correa, A., Cragan, J. D., Kucik, J. E., Alverson, C. J., Gilboa, S. M., Balakrishnan, R., ... Chitra, J. (2007). Metropolitan Atlanta Congenital Defects Program, 40th anniversary edition surveillance report. *Birth Defects Research, Part A: Clinical and Molecular Teratology, 79* (2), 1−120.

23. Rasmussen, S., Olney, R., Holmes, L., Lin, A., Keppler-Noreuil, K., Moore, C., & the National Birth Defects Prevention Study. (2003). Guidelines for case classification for the National Birth Defects Prevention Study. *Birth Defects Research, Part A: Clinical and Molecular Teratology, 67,* 193−201.

24. March of Dimes. (2011). *Birth defects.* Retrieved from <https://www.marchofdimes.org/complications/birthdefects.aspx>.

25. Brent, R. L. (1995). Bendectin: Review of the medical literature of a comprehensively studied human nonteratogen and the most prevalent tortogen−litigen. *Reproductive Toxicology Review, 9*(4), 337−349.

26. Brent, R. L. (2007). How does a physician avoid prescribing drugs and medical procedures that have reproductive and developmental risks? *Clinics in Perinatology, 34,* 233−232.

27. Brody, J. (1983, June 19). Shadow of doubt wipes out Bendectin. *New York Times.*

28. Wing, D. A., Powers, B., & Hickok, D. (2010). U.S. Food and Drug Administration drug approval: Slow advances in obstetric care in the United States. *Obstetrics & Gynecology, 115*(4), 825−833.

29. Merton, V. (1993). The exclusion of pregnant, pregnable, and once-pregnable people (a.k.a. women) from biomedical research. *American Journal of Law & Medicine, 19*(4), 369−451.

30. Baylis, F. (2010). Opinion: Pregnant women deserve better. *Nature, 465,* 689−670.

31. Mattison, D., & Zajicek, A. (2006). Gaps in knowledge in treating pregnant women. *Gender Medicine, 3*(3), 169−182.

32. Feibus, K., & Goldkind, S. F. (2011, May 17). Pregnant women and clinical trials: Scientific, regulatory, and ethical considerations. In *Oral presentation at the pregnancy and prescription medication use symposium.* Silver Springs, MD.

33. Greenwood, K. (2010). The mysteries of pregnancy: The role of law in solving the problem of unknown but knowable maternal-fetal medication risk. *University of Cincinnati Law Review, 79,* 267–322.

34. Brent, R. L. (2004). Utilization of animal studies to determine the effects and human risks of environmental toxicants. *Pediatrics, 113,* 984–995.

35. Brent, R. L. (2007). How does a physician avoid prescribing drugs and medical procedures that have reproductive and developmental risks? *Clinics in Perinatology, 34,* 233–232.

36. Charo, R. A. (1993). Protecting us to death: women, pregnancy, and clinical research trials. *Saint Louis University Law Journal, 38,* 135–187.

37. Eisenberg, R. S. (2005). The problem of new uses. *Yale Journal of Health Policy, Law, & Ethics, 5,* 717–718.

38. Eisenberg, R. S. (2005). The problem of new uses. *Yale Journal of Health Policy, Law, & Ethics, 5,* 717–718.

39. Kaposy, C., & Baylis, F. (2011). The Common Rule, pregnant women, and research: No need to "rescue" that which should be revised. *The American Journal of Bioethics, 11*(5), 60–62.

40. Merton, V. (1993). The exclusion of pregnant, pregnable, and once-pregnable people (a.k.a. women) from biomedical research. *American Journal of Law & Medicine, 19*(4), 369–451.

41. U.S. Government Printing Office. (2016, October 1). Code of Federal Regulations, title 45, part 46, subpart B. In *U.S. Government Printing Office via GPO Access* (pp. 140–143). Retrieved from <https://www.gpo.gov/fdsys/pkg/CFR-2016-title45-vol1/pdf/CFR-2016-title45-vol1-part46.pdf>.

42. Baylis, F. (2010). Opinion: Pregnant women deserve better. *Nature, 465,* 689–670.

43. Kaposy, C., & Baylis, F. (2011). The Common Rule, pregnant women, and research: No need to "rescue" that which should be revised. *The American Journal of Bioethics, 11*(5), 60–62.

44. Baylis, F. (2010). Opinion: Pregnant women deserve better. *Nature, 465,* 689–670.

45. McCullough, L. B., Coverdale, J. H., & Chervenak, F. A. (2005). A comprehensive ethical framework for responsibly designing and conducting pharmacologic research that involves pregnant women. *American Journal of Obstetrics and Gynecology, 193,* 901–907.

46. American College of Obstetricians and Gynecologists. (2015). ACOG Committee Opinion No. 646: Ethical considerations for including women as research participants. *Obstetrics & Gynecology, 126,* e100–e107. https://doi.org/10.1097/AOG.0000000000001150.

47. Greenwood, K. (2010). The mysteries of pregnancy: The role of law in solving the problem of unknown but knowable maternal-fetal medication risk. *University of Cincinnati Law Review, 79,* 267–322.

48. Feibus, K., & Goldkind, S. F. (2011, May 17). Pregnant women and clinical trials: Scientific, regulatory, and ethical considerations. In *Oral presentation at the pregnancy and prescription medication use symposium.* Silver Springs, MD.

49. Greenwood, K. (2010). The mysteries of pregnancy: The role of law in solving the problem of unknown but knowable maternal-fetal medication risk. *University of Cincinnati Law Review, 79,* 267–322.

50. Lyerly, A. D., Little, M. O., & Faden, R. (2008). The second wave: Toward responsible inclusion of pregnant women in research. *International Journal of Feminist Approaches in Bioethics, 1*(2), 5–22.

51. Baylis, F. (2010). Opinion: Pregnant women deserve better. *Nature, 465,* 689–670.

52. Lyerly, A. D., Little, M. O., & Faden, R. (2008). The second wave: Toward responsible inclusion of pregnant women in research. *International Journal of Feminist Approaches in Bioethics, 1*(2), 5–22.

53. Lyerly, A. D., Little, M. O., & Faden, R. (2008). The second wave: Toward responsible inclusion of pregnant women in research. *International Journal of Feminist Approaches in Bioethics, 1*(2), 5–22.

54. Merton, V. (1993). The exclusion of pregnant, pregnable, and once-pregnable people (a.k.a. women) from biomedical research. *American Journal of Law & Medicine, 19*(4), 369–451.
55. Kass, N. E., Taylor, H. A., & Anderson, J. (2000). Treatment of human immunodeficiency virus during pregnancy: The shift from an exclusive focus on fetal protection to a more balanced approach. *American Journal of Obstetrics and Gynecology, 182*(4), 1–5.
56. Brent, R. L. (2007). How does a physician avoid prescribing drugs and medical procedures that have reproductive and developmental risks? *Clinics in Perinatology, 34,* 233–232.
57. Kass, N. E., Taylor, H. A., & Anderson, J. (2000). Treatment of human immunodeficiency virus during pregnancy: The shift from an exclusive focus on fetal protection to a more balanced approach. *American Journal of Obstetrics and Gynecology, 182*(4), 1–5.
58. Lott, J. P. (2005). Module three: Vulnerable/special participant populations. *Developing World Bioethics, 5*(1), 30–53.
59. Center for Disease Control and Prevention. (2014). *National vital statistics report: International comparisons of infant mortality and related factors: United States and Europe, 2010.* Retrieved from <https://www.cdc.gov/nchs/data/nvsr/nvsr63/nvsr63_05.pdf>.
60. U.S. Central Intelligence Agency. (2017). *The World Factbook: Country comparison: Infant mortality rate.* Retrieved from <https://www.cia.gov/library/publications/the-worldfactbook/rankorder/2091rank.html>.
61. Kochanek, K. D., Murphy, S. L., Xu, J., & Arias, E. (2017). Mortality in the United States, 2016. In *NCHS data brief no. 293.* Retrieved from <https://www.cdc.gov/nchs/products/databriefs/db293.htm>.
62. Center for Disease Control and Prevention. (2017). *National Center for Health Statistics: Infant health.* Retrieved from <https://www.cdc.gov/nchs/fastats/infant-health.htm>.
63. Lyerly, A. D., Little, M. O., & Faden, R. (2008). The second wave: Toward responsible inclusion of pregnant women in research. *International Journal of Feminist Approaches in Bioethics, 1*(2), 5–22. https://doi.org/10.3138/ijfab.1.2.5.
64. Lyerly, A. D., Little, M. O., & Faden, R. R. (2011). Reframing the framework: Toward fair inclusion of pregnant women as participants in research. *The American Journal of Bioethics, 11*(5), 50–52. https://doi.org/10.1080/15265161.2011.560353.
65. Lyerly, A. D., Little, M. O., & Faden, R. (2008). The second wave: Toward responsible inclusion of pregnant women in research. *International Journal of Feminist Approaches in Bioethics, 1*(2), 5–22. https://doi.org/10.3138/ijfab.1.2.5.
66. Iber, F. L., Riley, W. A., & Murray, P. J. (1987). *Conducting Clinical Trials,* New York: Plenum Press. In Merton, V. (1993). The exclusion of pregnant, pregnable, and once-pregnable people (a.k.a. women) from biomedical research. *American Journal of Law & Medicine, 19*(4), 369–451.
67. Lupton, M. G. F., & Williams, D. J. (2004). The ethics of research in pregnant women: Is maternal consent sufficient? *British Journal of Obstetrics and Gynecology, 111*(12), 1307–1312. https://doi.org/10.1111/j.1471-0528.2004.00342.x.
68. Cain, J., Lowell, J., Thorndyke, L., & Localio, A. R. (2000). Contraceptive requirements for clinical research. *Obstetrics & Gynecology, 95*(6), 861–866. https://doi.org/10.2337/dc15-2723.
69. Edgar, H., & Rothman, D. J. (1990). New rules for new drugs: The challenge of AIDS to the regulatory process. *Milbank Quarterly, 68,* 111–114.
70. Macklin, R. (2010). The art of medicine: Enrolling pregnant women in biomedical research. *The Lancet, 375,* 632–633. https://doi.org/10.1016/S0140-6736(10)60257-7.
71. Macklin, R. (2010). The art of medicine: Enrolling pregnant women in biomedical research. *The Lancet, 375,* 632–633. https://doi.org/10.1016/S0140-6736(10)60257-7.
72. Berlin, J. A., & Ellenberg, S. S. (2009). Commentary: Inclusion of women in clinical trials. *BMC Medicine, 7*(56), 1–3. https://doi.org/10.1186/1741-7015-7-56.
73. Hall, J. K. (1995). Exclusion of pregnant women from research protocols: Unethical and illegal. *IRB: Ethics and Human Research, 17*(2), 1–3.

74. Merton, V. (1993). The exclusion of pregnant, pregnable, and once-pregnable people (a.k.a. women) from biomedical research. *American Journal of Law & Medicine, 19*(4), 369–451.

75. Merton, V. (1993). The exclusion of pregnant, pregnable, and once-pregnable people (a.k.a. women) from biomedical research. *American Journal of Law & Medicine, 19*(4), 369–451.

76. Beran, R. G. (2006). The ethics of excluding women who become pregnant while participating in clinical trials of anti-epileptic medications. *Seizure, 15,* 563–570. https://doi.org/10.1016/j.seizure.2006.08.008.

77. Macklin, R. (2010). The art of medicine: Enrolling pregnant women in biomedical research. *The Lancet, 375,* 632–633. https://doi.org/10.1016/S0140-6736(10)60257-7.

78. World Medical Association. (2008). *Declaration of Helsinki.* Retrieved from <https://www.wma.net/wp-content/uploads/2016/11/DoH-Oct2008.pdf>.

79. Kaposy, C., & Baylis, F. (2011). The Common Rule, pregnant women, and research: No need to "rescue" that which should be revised. *The American Journal of Bioethics, 11*(5), 60–62. https://doi.org/10.1080/15265161.2011.560360.

80. Hall, J. K. (1995). Exclusion of pregnant women from research protocols: Unethical and illegal. *IRB: Ethics and Human Research, 17*(2), 1–3.

81. Merkatz, R. (1998). Inclusion of women in clinical trials: A historical overview of scientific, ethical, and legal issues. *Journal of Obstetrical, Gynecologic, and Newborn Nursing, 27*(1), 78–84. https://doi.org/10.1111/j.1552-6909.1998.tb02594.x.

82. Institute of Medicine Committee on Women's Health Research. (2010). *Women's Health research: Progress, pitfalls, and promise.* Washington, DC: National Academies Press. Retrieved from <www.nap.edu>.

83. Institute of Medicine, Mastroianni, A. C., Faden, R., & Federman, D. (Eds.). (1994). *Women and health research: Ethical and legal issues of including women in clinical studies.* Washington, DC: National Academy Press.

84. Lyerly, A. D., Little, M. O., & Faden, R. R. (2011). Reframing the framework: Toward fair inclusion of pregnant women as participants in research. *The American Journal of Bioethics, 11*(5), 50–52. https://doi.org/10.1080/15265161.2011.560353.

85. Merton, V. (1993). The exclusion of pregnant, pregnable, and once-pregnable people (a.k.a. women) from biomedical research. *American Journal of Law & Medicine, 19*(4), 369–451.

86. McCullough, L. B., Coverdale, J. H., & Chervenak, F. A. (2005). A comprehensive ethical framework for responsibly designing and conducting pharmacologic research that involves pregnant women. *American Journal of Obstetrics and Gynecology, 193,* 901–907. https://doi.org/10.1016/j.ajog.2005.06.020.

87. Hall, J. K. (1995). Exclusion of pregnant women from research protocols: Unethical and illegal. *IRB: Ethics and Human Research, 17*(2), 1–3. https://doi.org/10.2307/3563527.

88. Hall, J. K. (1995). Exclusion of pregnant women from research protocols: Unethical and illegal. *IRB: Ethics and Human Research, 17*(2), 1–3. https://doi.org/10.2307/3563527.

89. Merton, V. (1993). The exclusion of pregnant, pregnable, and once-pregnable people (a.k.a. women) from biomedical research. *American Journal of Law & Medicine, 19*(4), 369–451.

90. Frank, E., & Novick, D. M. (2003). Beyond the question of placebo controls: Ethical issues in psychopharmacological drug studies. *Psychopharmacology, 171*(1), 19–26. https://doi.org/10.1007/s00213-003-1477-z.

91. Institute of Medicine, Mastroianni, A. C., Faden, R., & Federman, D. (Eds.). (1994). *Women and health research: Ethical and legal issues of including women in clinical studies.* Washington, DC: National Academy Press.

92. Lyerly, A. D., Little, M. O., & Faden, R. (2008). The second wave: Toward responsible inclusion of pregnant women in research. *International Journal of Feminist Approaches in Bioethics, 1*(2), 5–22. https://doi.org/10.3138/ijfab.1.2.5.

93. Macklin, R. (2010). The art of medicine: Enrolling pregnant women in biomedical research. *The Lancet, 375,* 632–633. https://doi.org/10.1016/S0140-6736(10)60257-7.

94. American College of Obstetricians and Gynecologists. (2015). ACOG Committee Opinion No. 646: Ethical considerations for including women as research participants. *Obstetrics & Gynecology, 126*, e100–e107. https://doi.org/10.1097/AOG.0000000000001150.

95. Institute of Medicine, Mastroianni, A. C., Faden, R., & Federman, D. (Eds.). (1994). *Women and health research: Ethical and legal issues of including women in clinical studies.* Washington, DC: National Academy Press.

96. Council for International Organizations of Medical Sciences. (2016). *CIOMS international ethical guidelines for health-related research involving humans.* Retrieved from <https://cioms.ch/wp-content/uploads/2017/01/WEB-CIOMS-EthicalGuidelines.pdf>.

97. Feibus, K., & Goldkind, S. F. (2011, May 17). Pregnant women and clinical trials: Scientific, regulatory, and ethical considerations. In *Oral presentation at the pregnancy and prescription medication use symposium.* Silver Springs, MD.

98. Macklin, R. (2010). The art of medicine: Enrolling pregnant women in biomedical research. *The Lancet, 375*, 632–633. https://doi.org/10.1016/S0140-6736(10)60257-7.

99. McCullough, L. B., Coverdale, J. H., & Chervenak, F. A. (2005). A comprehensive ethical framework for responsibly designing and conducting pharmacologic research that involves pregnant women. *American Journal of Obstetrics and Gynecology, 193*, 901–907. https://doi.org/10.1016/j.ajog.2005.06.020.

100. Mohanna, K., & Tunna, K. (1999). Withholding consent to participate in clinical trials: Decisions of pregnant women. *British Journal of Obstetrics and Gynaecology, 106*, 892–897. https://doi.org/10.1111/j.1471-0528.1999.tb08426.x.

101. Mohanna, K., & Tunna, K. (1999). Withholding consent to participate in clinical trials: Decisions of pregnant women. *British Journal of Obstetrics and Gynaecology, 106*, 892–897, citing Foster, C., (Ed.). (1997). *Manual for research ethics committees.* London: Centre of Medical Law and Ethics.

102. Lyerly, A. D., Little, M. O., & Faden, R. (2008). The second wave: Toward responsible inclusion of pregnant women in research. *International Journal of Feminist Approaches in Bioethics, 1*(2), 5–22. https://doi.org/10.3138/ijfab.1.2.5.

Chapter 3

The ethics involved

Theoretical approaches

Conceptual or theoretical frameworks can be described as a group of concepts that are broadly defined and systematically organized to provide a focus, a rationale, and a tool for the integration and interpretation of information.[1]

In this area of inquiry, where both the pros and cons of the inclusion of pregnant women in research can be argued on the basis of ethical considerations, an ethical framework can assist to provide structure to the debate. A common language that is familiar to and accepted by health care practitioners, health care researchers, pregnant women, academics, and pharmaceutical industry researchers, will assist stakeholders in the understanding and consideration of ideas, opinions, and options. The proposed ethical framework of clinical research is based upon moral reasoning and ethical concepts that are common to these stakeholders' histories, lived experiences, and values at home and in the workplace.

Medical practice and clinical research are closely related and share Hippocratic roots, although important differences have been noted. Medical practice focuses on the improvement of an individual's health and well-being, while clinical research attempts to provide knowledge that will improve a population's health and well-being by identifying improved methods to treat, cure, or prevent disease.

Medical ethics and research ethics are also closely related and many medical practitioners participate in clinical research. Largent et al. cite technological advances such as electronic medical records, increased demand for evidence-based medicine and comparative effectiveness research, and the recognition that participation in research allows access to new therapies during "evidence development" as reasons for an increased blurring of clear boundaries between research and the provision of care.[2] It needs to be recognized by both the researcher and the participant that, at times, the aims of the research may not coincide with the individual interests of the participant.

"The fundamental ethical concern raised by clinical research is whether and when it can be acceptable to expose some individuals to risks and burdens for the benefit of others. A full analysis of the ethics of exposing subjects to risks needs to justify both the treatment of the subjects and the

Pregnancy and the Pharmaceutical Industry. DOI: https://doi.org/10.1016/B978-0-12-818550-6.00003-9

behavior of the researchers."[3] To address these issues, a brief review of some of the most common theories in the field of research ethics is presented below.

Principle-based ethics

Principlism is the ethical approach traditionally applied to the fields of clinical practice, medical research, epidemiology, and public health. In 1979, American ethicists Tom Beauchamp and James Childress published the Principle of Biomedical Ethics[4] in which they described four principles: respect for autonomy, beneficence, nonmaleficence, and justice, as the basis for resolving ethical issues in clinical medicine. In the same year, the Belmont Report[5] identified respect for persons, beneficence, and justice as the basis for resolving ethical issues in research using human subjects.[6] And thus, since the mid-1900s, clinical and research ethics have been linked. Principlism is sometimes criticized as being too "Western," lacking sensitivity to other cultural perspectives and too theoretical and broad to be applied to widely variable specific situations.[7] However, it remains the dominant theory applied to clinical practice and research.

The principle of respect for autonomy or respect for persons

Twentieth century legal decisions have determined that the authority and power to authorize a health care provider (HCP) to act on a patient's behalf is vested in the adult, competent patient. HCPs have the obligation to explain their rationale for recommending an intervention, the risks and benefits of the proposed procedures, alternative interventions they are not recommending, and the reasoning behind these recommendations. The competent adult patient is solely responsible for authorizing the initiation of the intervention.

Each patient brings her "unique configuration and history of particular values and beliefs that form the basis for [the] determination of [their] own subjective and deliberative interests"[8] to the situation. Thus the patient is the only one who can make decisions that are relevant in her own context.

But with the complexity and multitude of options in today's medical environment, this deciding cannot occur in a vacuum. Four models of increasingly respectful doctor-assisted patient decision-making were described by Emanuel in 1992: paternalistic, informative, interpretative, and deliberative.[9] In the preferred deliberative model, the caring HCP teaches the patient about her medical condition and treatment options, discusses both in the context of the values held by the HCP and by the patient, and provides a recommendation based on these factors. In today's environment of overwhelming access to medical information and misinformation, it is imperative that medical providers share perspective as well as knowledge. Kukla characterizes the exchange of knowledge and context by the patient and HCP as active collaborative knowledge-building.[10] With this assistance, the patient can then make her informed decision based on the consideration of information and value context. Thus, respect for autonomy means more than just having the patient make the decision. Today it means respecting the patient's capacity for learning, for evaluating values and context, and for applying these considerations to arrive at a considered decision.

The same principles invoked in the therapeutic environment hold true in the research environment. According to McCullough,[11] the principle of autonomy is a construct of modern Western political philosophy. It is the first principle of the Nuremberg Code,[12] and the Belmont Report.[13] In practical terms, one way of ensuring autonomous choice had been through the provision of informed consent. Informed consent can be defined as "consent given by a competent person in the light of relevant information and without the presence of any pressure or coercion."[14] Its application to research has been influenced and upheld by legal proceedings that address the independent moral status of people and respect for their self-determination.

The principles of nonmaleficence, beneficence, and the double effect

Though widely quoted and attributed to Hippocrates, the edict, "primum non nocere (first, do no harm)" does not occur in the Hippocratic Oath. A similar statement, "make a habit of two things — to help, or at least to do no harm" occurs in an accompanying text and combines the principles of beneficence—to do good, and nonmaleficence—to do no harm. Nonmaleficence is, according to some ethicists, a corollary of the principle of beneficence and not an independent principle on its own.[15] The meaning and import of both principles are obvious: the physician commits to seek interventions on the patient's behalf, the consequences of which are intended to provide benefit and not cause harm. Adherence to these principles in practice encourages the practitioner to utilize her "accumulated scientific and clinical knowledge, skill, and experience" to "protect and promote the interests of the patient."[16] Thus the strength of an HCP's beneficence is limited by the "competencies"[17] of the HCP's medical judgment at any point in time.

In medical and public health practice there are often situations where an action intended to provide benefit may result in an inadvertent harm—the harm being intentionally less likely to occur or of an acceptable quality or intensity. Examples include mandatory vaccinations that cause adverse effects in some children but benefit many, or surgery to remove an ectopic pregnancy that saves the woman's life but ends the pregnancy. Patients weigh potential benefits against the risk of potential adverse effects when choosing to take medication or consenting to surgery.

The double effect principle recognizes the artificial separation of beneficence from nonmaleficence. It provides criteria to help judge if an intervention is morally acceptable including if (1) the action is morally neutral or good, (2) the intention is to invoke the good effect and not the bad one, (3) the situation is serious enough that the harmful side effect is justifiable, and (4) actions are taken to minimize the potential harmful effects. Originated by St. Thomas Aquinas in his deliberations about the morality of killing in self-defense, the double effect principle is applied in law, medicine, research, and the military to evaluate the ethics of actions with good intentions but adverse consequences.[18] Complicating the matter further, whether an outcome is a benefit (or a harm) can be judged differently by different people or by the same person in different contexts. Pregnancy itself is an example of such a dependent outcome—a joyfully anticipated occurrence at some times, a burden or even a danger to health and well-being in other people, contexts, or times. This lack of objective yardstick with which to assess the morality of any decision further necessitates fully informed consent and respect for autonomy.

In research, the principles of beneficence, nonmaleficence, and the double effect are operationalized in study design, methodology, and protocol development that minimize risks to the study participants. An Institutional Review Board (IRB), also called an independent ethics committee or an ethical review board, is responsible for objectively evaluating each study for its adherence to these principles, to prospectively assess the study's risks and benefits, and to assure potential harms to participants have been minimized. In the United States, an IRB review is required by the Food and Drug Administration (FDA) and the Office for Human Research Protection for any research that receives support either directly or indirectly from the Department of Health and Human Services.[19] IRBs are guided by (1) The Common Rule [the common name for Subpart A, Part 46: Protection of Human Subjects, of Title 45: Public Welfare, in the Code of Federal Regulations (46CFR45) which contains the basic policy for protection of human research subjects] and (2) the international guidance documents that address human subjects research (e.g., the Nuremberg Code, the Declaration of Helsinki, the Belmont Report, and CIOMS guidelines.)[20]

The principle of justice

Justice was the "guiding ethical principle for the IOM committee" in their evaluation of the inclusion of women in clinical research.[21] The Belmont Report discusses the principle of justice in the research setting as requiring an "equitable distribution of the burdens and benefits of research."[22] Researchers must not include without good reason eligible candidates who may be harmed by participation, that is, vulnerable persons. Nor can researchers "exclude without good reason eligible candidates who may benefit from participation."[23] Some would argue that, when pregnant women do not have access to clinical research studies, they are denied the possibility of having the potential for benefit that is available to nonpregnant women. This would violate the principle of justice.[24]

Many individuals have been excluded from research as a means to protect them from being unfairly burdened by the potential harms of research; some are categorized in federal guidelines as "vulnerable." Yet Levine et al. warn that the "concept of vulnerability stereotypes whole categories of individuals," including pregnant women, as being less than capable of making reasoned decisions.[25] The label itself may lead to injustice because it results in the exclusion of people who may indeed have robust decisional capability from the opportunity to exercise that capacity in deciding whether their participation in research is in their best interests. Justice, state Lyerly et al., "calls into question the de facto summary exclusion of pregnant women in research without justification."[26]

Consequentialism

The question of inclusion/exclusion is at the basis of this issue. In addition to Principlism, another "action-based" approach to ethics that addresses this concern is Consequentialism. This theory holds that the moral status of an action is determined by the goodness or badness of its outcomes. The Declaration of Helsinki, one of the most influential documents addressing clinical research, includes, in addition to autonomy, beneficence, and justice, an additional moral requirement for ethical conduct: that patients' participation in research should not put them at a disadvantage with respect to medical care. One could argue that this requirement includes and requires its converse: that patients' *exclusion* from research should not put them at a disadvantage with respect to medical care. Stakeholders argue that the exclusion of pregnant women from medical research puts them decidedly at such a disadvantage. States Lisa Eckenwiler in a paper entitled, Hopes for Helsinki: Reconsidering vulnerability, "...in this contemporary era of research, it is essential that codes of ethics move beyond merely protectionist thinking. Fair access to research participation should be addressed more explicitly."[27]

Eckenweiler takes the access issue a step further and thinks we should "...extend the scope of responsibility for ethical research to industry leaders, elected officials, and research funders, because they too play a role in ensuring that research endeavours do not create or perpetuate vulnerabilities, particularly inequalities in health or relations of power."[28] Inequalities of power are further addressed by the feminist theories that developed in the mid-20th century.

Feminist ethical theory

The modern feminist approaches to ethical analysis arose in response to dissatisfaction with the implications of principlistic "moral values."[29]

Traditional ethics were criticized for promoting culturally masculine traits like "independence, autonomy, intellect, will, hierarchy, domination, culture, wariness, war, and death" while dismissing culturally feminine traits like "interdependence, community, emotion, sharing, absence of hierarchy, connection, nature, trust, peace, and life."[30] Traditional ethics were also said to favor "'male' ways of moral reasoning that emphasize rules, rights, universality, and impartiality over 'female' ways of moral reasoning that emphasize relationships, responsibilities, particularity, and partiality."[31] Feminist perspectives sought to promote the importance of subjective experience in moral reasoning.

Feminist ethics is often linked with the "ethic of care." A term coined by moral psychologist Carol Gilligan, the ethic of care highlights certain salient moral considerations (context, particularity, relationships) that have not received due attention from traditional moral theories. But the ethic of care is only one of many feminist approaches. Others include liberal, radical, Marxist/socialist, multicultural, and ecological ethics where the emphasis is on questions of internal and external power, domination, and subordination. Existentialist, postmodern, and Third-Wave approaches focus on the psychological consequences of social status. These feminist approaches seek to identify and address the ways in which gender, class, and culture affect moral decision-making.

The feminist ethic of care aims primarily to identify and improve women's conditions—and, by extension, to improve circumstances for other vulnerable people like children, the elderly, and minorities.[32] Its tenets include loving, caring, empathy, sensitivity, and an emphasis on relationships and responsibilities. As the topic of this inquiry is the pregnant woman's relationships and responsibilities within the research community, with her HCP and, most importantly, with her fetus, the care-oriented approach is highly relevant. The feminist ethic of care is consistent with, but expands, the principlist ethic of beneficence.

The emphasis on relationships and responsibilities is in contrast to the individualistic approach in traditional ethics which are concerned with the rights of the individual. Carol Gilligan described moral development as growth from the individualistic perspective of the infant and child, to the developing realization of the person in relation to others, and finally to the mature person who can balance her individual needs with the needs of others.[33] This moral maturity results in an acceptance of responsibility for oneself and for the effect of one's actions on others.

In this case, one might see the inclusion/exclusion question about pregnant women in clinical research from a traditional (i.e., male) perspective: avoiding harm by applying a rule (de jure or de facto) to all clinical studies; no pregnant women (or women capable of becoming pregnant) in clinical trials, period. But Gilligan, citing Piaget's findings, found that the female perspective encompassed "a greater tolerance, a greater tendency toward innovation in solving conflicts, a greater willingness to make exceptions to rules, and a lesser concern with legal elaboration."[34] Thus, including the feminist ethic in the framework of this field of inquiry not only promotes the perspective of the primary subject of concern (women), but broadens the discussion and opens the deliberation to innovation, flexibility, context, and particularity.

Rather than being dichotomous or contradictory, however, can we find parallels between the traditionalist and the feminist approaches? The principles of autonomy, beneficence, and justice are the foundation of medical and research ethics both in theory (see the ethical codes) and in clinical and research practice. Are these principles compatible with feminist theory and practice?

Paternalism—"overriding the [competent person's] wishes or intentional actions [even if] for beneficent reasons"[35]—is an "offense to the autonomy of [competent] persons"[36] and would be unacceptable to the principlist's "respect for persons" and to the "ethic of care." Beneficent, care-oriented practitioners would embody respect for persons by sharing their knowledge, their experience, their preferences, and the rationale for those preferences with the woman/patient/potential research subject. They would encourage the woman to include her own unique subjective experiences, preferences, and knowledge interpretations in her deliberations. Perspective sharing from the woman's own "experts" of choice—her personal supporters (e.g., the father of the fetus, family members, trusted friends)—would also be encouraged. The formulation of a decision in this manner is dense with relationships and responsibilities. Thus, the care-oriented practitioner, in respectful relationship with her patient, enables her autonomy.[37,38] This is fully informed consent.

Parallels exist between the ethic of care and the principle of beneficence because both seek to induce beneficial outcomes and refrain from causing harm. Nel Noddings suggests that the caring behavior we naturally exhibit as children, helping others simply because we want to help them, develops into ethical caring as we grow to live in the complex external world.[39] She further states that "having a robust sense of social justice is predicated on the lessons learned in the private sphere." Thus there are parallels between beneficence and justice and the ethic of care perspective as well. This suggests that both the principle-based approach and the care-oriented approach can, and do, coexist. Carol Gilligan states that all human relationships can be viewed from the justice perspective in terms of equality ("don't act unfairly towards others") and from the ethic of care perspective in terms of attachment ("don't turn away from someone in need").[40] Utilizing both perspectives buttresses the argument for inclusiveness. Gilligan challenges both men and women to "speak the moral language of justice and rights as fluently as the moral language of care and responsibility."[41]

In conjunction with the principle-based ethical framework that is commonly evoked by medical and research practitioners, the feminist ethic of care perspective is relevant to this issue. Caring is central to women's experience—not just pregnant women's experience. Women may be never pregnant, prepregnant, pregnant, or postpregnant at different times in their lives. And many women and men care for and about pregnant women—be they their HCPs, their significant others, their parents, their children, or their friends. The safety and effectiveness of medical intervention in clinically compromised pregnancies impact, and is of concern to, many people on a fundamentally sensitive, loving, and caring basis.

Alison Jaggar[42] describes the outcomes that all approaches to feminist ethics seek to achieve:

- To articulate moral critiques of actions and practices that perpetuate women's subordination.
- To prescribe morally justifiable ways of resisting such actions and practices.
- To envision morally desirable alternatives for such actions and practices.
- To take women's moral experience seriously, though not uncritically.

These aims parallel the aims of our inquiry: in articulating the views of the US pharmaceutical industry, to find common purpose and mutually beneficial approaches to including pregnant women who have historically been excluded from clinical research studies.

Business ethics

The discovery and development of pharmaceutical agents is a complex and costly endeavor. Knowledge of chemistry, pharmacology, biochemistry, microbiology, genomics, toxicology, and clinical medicine are required. Once a potentially viable product reaches the end of the labyrinth of development, the product then needs manufacture, packaging, production, distribution, promotion, ongoing monitoring, fiduciary, and legal support. The need for all of these highly sophisticated and closely regulated functions requires a complex interdisciplinary organization—the pharmaceutical industry.

In the pharmaceutical industry, medical practitioners work collaboratively with research scientists on the design, conduct, and evaluation of clinical studies. The goal to obtain objective and definitive scientific data that support the use of preventative or curative therapies by people in need satisfies both the clinical and the research agenda. But there is a third agent whose needs are also present—the corporate agenda.

What are the generally acceptable ethical principles for the corporation? Many articles analyzing ethical failures were available but I was not able to identify any one salient, widely-accepted ethical framework for business. I did find an instructive description of three potential vantage points from which business ethics could be derived.[43] They are:

- Ethics derived from the profit motive—including "good ethics result in good business," that is, the best interests of a business are served by establishing a trusting relationship with the public (resulting in increased product loyalty, and decreased liability claims) and employees (resulting in good morale and productivity) and its reverse, "good business results in good ethics," that is, the demand for moral behavior from customers and employees will result in proper behavior from companies. Businesses that meet these demands will survive and prosper.

- Ethics derived from the legal system—businesses will do the right thing as prescribed by the law; any obligation beyond the law is optional.
- Ethics derived from general moral obligations—
  - *Harm principle*: businesses should avoid causing unwarranted harm.
  - *Fairness principle*: business should be fair in all of their practices.
  - *Human rights principle*: businesses should respect human rights.
  - *Autonomy principle*: businesses should not infringe on the rationally reflective choices of people.
  - *Veracity principle*: businesses should not be deceptive in their practices.
  - *Stakeholder principle*: businesses should consider all stakeholders' interests that are affected by a business practice.

Most companies utilize a combination of these three approaches to guide decision-making, and the pharmaceutical companies would be no exception. But medical research is fundamental to this industry's ability to discover, produce, and maintain its products. Therefore the ethical framework for medical research must be compatible with the ethical framework of the business. The third approach, ethics derived from general moral obligations, is the most compatible with research ethics and the principle-based and feminist approaches described above. The Harm and Veracity principles correspond to Nonmaleficence, the Fairness principle to Justice, and the Autonomy and Human rights principles to respect for person.

One could propose that the Stakeholder principle corresponds to the consequentialist principle discussed above, that is, the business practice of excluding pregnant women from research studies may not be in women's best interests. On the other hand, including pregnant women in research studies may not be in the company's best interests.

The founders of Stakeholder theory, Freeman and Gilbert,[44] "view business as a connected set of relationships among stakeholders built [not upon the traditionalist] principles of competition and justice but [on] cooperation and caring."[45] Burton and Dunn[46] propose parallels between stakeholder and feminist theories as these both promote the centrality of relationships as the basis upon which knowledge is gained, options are considered, and decisions should be made. In individualistic rights-based organizational theory, the firm is in competition with other firms and seeks to further its own interests.

Legal contracts replace trust-based relationships to protect the company in negotiations. In stakeholder theory, firms take a more cooperative stance, seeking decisions where all parties gain. This can only happen when the company makes the stakeholders' interests explicit and considers the effects of the proposed decisions on all parties involved. Burton and Dunn believe that a rights-based view is inherently problematic when differences of opinion arise. If competing parties are trying to get their "inherent rights" met— but no one's rights supersede another one's rights—a stalemate is inevitable. In Stakeholder theory, companies try to do the "right" thing. When differences of opinion arise, Burton and Dunn invoke the ethic of care and propose that the company should "care enough for the least advantaged stakeholders that they not be harmed. A firm following this principle must perform a stakeholder analysis not merely to understand which stakeholders have power and which have a stake in the decision, but also to understand which stakeholders are most vulnerable to the action."[47] This combination of stakeholder theory and feminist ethics may inform our conceptual approach to interactions with pharmaceutical company and related organization representatives and to the proposal of procedural change within the industry.

There are many examples of pharmaceutical company philanthropic activities contributing to the public health needs of populations at risk. Some feel that these "good deeds have not been given the credit and recognition they deserve."[48] They have been overshadowed by stories of unlawful marketing practices, unfair pricing practices, suppression of safety and efficacy data, and manipulative business practices.[49] The challenge before the pharmaceutical industry is to find the portfolio of products that maintains commercial success (ensuring profitability and shareholder value) while addressing a range of unmet medical needs (fulfilling social and political responsibilities). To do the right thing and to overcome negative public perception based on past shortcomings, the successful company must maintain corporate integrity by strictly adhering to ethical research and business practices.

Special considerations for pregnancy/maternal-fetal ethics

Vulnerable subjects are those people whose rights need the most protection. These groups, according to federal regulations, include children, incapable adults, prisoners, and pregnant women. Lupton states that, "While children, incapable adults, and prisoners are vulnerable because they lack freedom or autonomy, what distinguishes pregnant women from the rest of the population is the prospect of causing harm to vulnerable 'future people' (their unborn children)."[50] According to him, it is the fetus who is vulnerable to having its consent given by a proxy since it is incapable of giving it itself. Do we suspect that a pregnant woman might disregard or minimize the risk to her fetus when a study might provide benefit to her? Benefit to the health of the pregnant woman usually provides benefit to the fetus as well. Conversely, a medically compromised fetus can jeopardize the life of the pregnant woman. Benefit to either party usually provides benefit to both, as harm to either party may cause harm to both.

Or is the gravid woman vulnerable in her own right? A woman does not lose reasoning capacity when she becomes pregnant. Yet her concern for her developing fetus might influence her decision-making. Might a pregnant woman accept a higher risk to herself in the interests of aiding her fetus? Might it be a risk that would be unacceptable to a nonpregnant subject? Is this what the FDA research guidelines attempt to address by requiring, to the extent possible, that the father of the fetus also provide consent in the case of a therapeutic intervention that is beneficial only to the fetus? Is it to protect the woman from herself (her altruism, her love, her compassion)? Or is the regulatory agency implying that the woman is less than capable of making the decision herself?

Is the pregnant research subject more or equally prone to "therapeutic misconception"[51]—"the tendency of patient-subjects to mistakenly assume that research interventions are designed to benefit them"[52]—and thus be more willing to consent to participate in a study? Or are they more prone, as research has shown, to overestimate the risk of all exposures that inadvertently or purposefully occur during the course of a pregnancy[53,54] – and thus be less likely to participate in research?

Is the classification of the pregnant woman as "vulnerable" there to protect her interests or to protect the fetus' interests? Is it the pregnant woman who is vulnerable, or should we say that children, the incompetent, prisoners, and *fetuses* are vulnerable groups?

I find that I am not alone in my perplexity about the characterization of pregnant women as a vulnerable group. In a paper entitled "The two dimensions of subject vulnerability"[55] in the *Journal of Clinical Research Best Practices*, the author depicts vulnerable groups on a graph. On the *x*-axis is the "Ability to Give Informed Consent" and on the *y*-axis is "Resistance to Undue Influence and Coercion" with the vulnerable groups ranging in placement on scales from low to high on each axis. The 20 vulnerable groups depicted include, among others, all of the vulnerable groups defined in the Code of Federal Regulation—except pregnant women. It seems Mr. Goldfarb did not know where to place pregnant women either. He did, however, include "Unborn" as a vulnerable group, low on the x-axis and high on the *y*-axis. So perhaps he agrees with Lupton and interprets the regulations to mean that it is the fetus and not the pregnant woman who is vulnerable.

ACOG supports a recommendation made by the NIH Office of Research on Women's Health in 2010 to define pregnant women as "scientifically complex" rather than "vulnerable."[56] Other groups in the HHS "vulnerable" category may not have the ability to provide rationally informed consent, or may be vulnerable to coercion, neither of which apply to most pregnant women. ACOG Committee Opinion #646 describes pregnant women as complex both physiologically (changes in anatomy and physiology) and ethically (needing to balance multiple interests) due to the presence of the fetus.[57] This description better describes the pregnant woman's situation and may ease the transition away from the inaccurate, paternalistic, and consequential classification of "vulnerable."

Respect for persons requires that the choices of rational individuals are respected and that people who are incapable of making their own choices are protected. This presents a conundrum for the investigator with a pregnant subject. On the one hand, the pregnant woman and the fetus are not independent. The fetus is wholly dependent on its "host." Therefore the mother's decision should be sufficient consent. But as the fetus is incapable of making its own choice, protection is projected in his or her interest.

Who speaks for the fetus? According to HHS (and subsequently FDA) guidelines, it is sometimes the mother and sometimes the mother and the father. According to American jurisprudence, it is sometimes the government. There are cases of government ruling in favor of HCPs to force pregnant women to have C-sections for the "good of the fetus." In one notorious case, neither the mother nor the baby survived. In that case, the court ruled after the fact that "neither fetal rights nor state interests on behalf of the fetus supersede women's rights as ultimate medical decision maker."[58]

The ACOG Committee on Ethics has taken a position on the balancing of the mother's and the fetus' interests. The Ethics Committee Opinion #664 states that "The College opposes the use of coerced medical interventions for pregnant women, including the use of the courts to mandate medical interventions for unwilling patients." The opinion considers that court-ordered obstetric interventions have disproportionately affected disadvantaged women including women of color, women of low socioeconomic status, and women who did not speak English. Even more concerning are the results of a 1986 study finding that in "almost one third of cases in which court orders were sought, the medical judgment, in retrospect, was incorrect."[59]

ACOG recognizes that the interests of the maternal-fetal unit usually converge rather than diverge. However, even when they don't, the responsibility to consent or refuse to participate in medical therapy or in clinical research remains with the woman.[60]

Application of an ethical framework for studies with pregnant women

Having reviewed the ethical principles and theories most frequently invoked when discussing clinical research, medical practice, women's interests, and business practices, I concluded that an ethical framework for evaluating perspectives about the inclusion of pregnant women in clinical research should be based upon the ethical principles of:

- Autonomy, or Respect for Persons, including the avoidance of paternalism.

- Beneficence, Nonmaleficence, and the Double Principle, including the avoidance of causing disadvantage.
- Justice.
- Care.
- Stakeholder Considerations, including as stakeholders both the pregnant woman and her caregivers and the drug company and its researchers.

TABLE 3.1 Ethical guidance: application of ethical principles to proposed solutions.

Autonomy/Respect:

Does this rationale/solution impose on anyone's personal autonomy?

Do all relevant parties consent to this rationale/solution? If not, what are the objections? Are all opinions acknowledged and respected?

Beneficence:

Who benefits from this rationale/solution and in what way?

Does the rationale/solution use the best of our current knowledge?

Does the rationale/solution favor the balance of benefit over risk?

Nonmaleficence:

Who may be harmed by this rationale or the implementation of this solution?

How have the potential harms been minimized?

Are risks communicated in a truthful, complete, and open manner?

Justice:

Is the rationale/proposed solution equitable to all stakeholders?

Can it be made to be more equitable?

Are the benefits and the burdens fairly distributed among stakeholders?

Ethic of care:

Whose needs are being met by this solution?

Does the rationale/solution promote cooperation among stakeholders?

Are relationships identified and maintained or promoted by the action?

Stakeholders:

Have all parties involved in this rationale/solution been identified?

What parties are impacted by this rationale/solution and in what way?

Are all stakeholders' concerns respected and addressed?

Do they agree on the solution?

[a]*Beauchamp, T., & Childress, J. (2001). Principles of biomedical ethics (5th ed.). Oxford University Press.*
Source: Adapted from L. Carter, A Primer to Ethical Analysis, and Beauchamp & Childress, Principles of Biomedical Ethics.[a]

This ethical framework aided the construction of the interview guide I used by helping to identify the relative importance of potential questions and what terminology and concepts would be most readily appreciated by the interviewees. The ethical issues raised are best explored via dialogue—hence the interview approach.

This ethical framework may aid the application of ethical principles to potential solutions proposed to address the dilemma of whether or not to include pregnant women in research studies. As with most ethically laden situations, no completely "right" or "wrong" solutions will likely be obvious. No one solution will be correct in every situation. Just as it would be unethical to require the inclusion of pregnant women in all clinical trials, the exclusion of pregnant women from all clinical trials is not ethically permissible either. Having an ethical framework to inform deliberations will facilitate the formulation of solutions and exceptions to those solutions upon which stakeholders can agree. Table 3.1 illustrates a potential application of this ethical framework to the potential solutions proposed by various stakeholders to address the inclusion of pregnant women in clinical research. This framework provides a structure and a common basis upon which to justify conclusions. In this arena, where both the pros and cons of the inclusion of pregnant women in research can be argued on ethical merits (see Chapter 2: The rationales for and against inclusion), a framework in the accepted language of medical, research, feminist, and business theory, will assist stakeholders in the understanding and consideration of options.

Notes

1. *Conceptual framework. Mosby's medical dictionary, 8th edition.* (2009). From <https://medical-dictionary.thefreedictionary.com/conceptual + framework> Retrieved 16.10.18.
2. Largent, E. A., Joffe, S., & Miller, F. G. (2011). Can research and care be ethically integrated? *Hastings Center Report, 41*(4), 37–46, p. 38. <https://doi.org/10.1002/j.1552-146X.2011.tb00123.x>
3. Wendler, D. (2017, Winter). The ethics of clinical research. In E. Zalta (Ed.) *The Stanford Encyclopedia of Philosophy.* Retrieved from <https://plato.stanford.edu/entries/clinical-research/>.
4. Beauchamp, T., & Childress, J. (2013). *Principles of biomedical ethics* (7th ed.). New York: Oxford University Press. (1st ed. published in 1979).
5. *National Commission for the Protection of Human Subjects of Biomedical and Behavioral Research. The Belmont Report.* (1979). Retrieved from <https://www.hhs.gov/ohrp/regulations-and-policy/belmont-report/read-the-belmont-report/index.html>.
6. McCormick, T. R. (2013). *Principles of bioethics.* Ethics in Medicine, University of Washington School of Medicine. Retrieved from <https://depts.washington.edu/bioethx/tools/princpl.html>.
7. Langley, G. C., & Egan, A. (2012). The ethics of care in biomedical research committees. *Journal of Clinical Research and Bioethics, 3,* 128. <https://doi:10.4172/2155-9627.1000128>.
8. McCullough, L. B., & Chervenak, F. A. (1994). *Ethics in obstetrics and gynecology.* New York: Oxford University Press, Inc. <https://doi.org/10.1046/j.1469-0705.1995.05060424-2.x>.

9. Emanuel, E. J., & Emanuel, L. L. (1992). Four models of the physician-patient relationship. *Journal of the American Medical Association, 267*(16), 2221−2226. <https://doi.org/10.1001/jama.1992.03480160079038 >.

10. Kukla, R. (2007). How do patients know? *Hastings Center Report, 37*(5), 27−35. <https://doi.org/10.1353/hcr.2007.0074>.

11. McCullough, L. B., & Chervenak, F. A. (1994). *Ethics in obstetrics and gynecology.* New York: Oxford University Press, Inc. <https://doi.org/10.1046/j.1469-0705.1995.05060424-2.x>.

12. The Nuremberg Code. (1947). *Trials of War Criminals before the Nuremberg Military Tribunals under Control Council Law No. 10, Vol. 2, 181−182.* Washington, D.C.: U.S. Government Printing Office, 1949. Retrieved from <https://history.nih.gov/research/downloads/nuremberg.pdf>.

13. National Commission for the Protection of Human Subjects of Biomedical and Behavioral Research. (1979). *The Belmont Report.* Retrieved from <https://www.hhs.gov/ohrp/regulations-and-policy/belmont-report/read-the-belmont-report/index.html>.

14. Spencer, A. S., & Dawson, A. (2004). Implications of informed consent for obstetric research. *The Obstetrician & Gynaecologist, 6,* 163−167. 10.1576/toag.6.3.163.26999, citing Doyal, L. (1997). Informed consent in medical research. *British Medical Journal, 314,* 1107−1111.

15. McCullough, L. B., & Chervenak, F. A. (1994). *Ethics in obstetrics and gynecology.* New York: Oxford University Press, Inc. <https://doi.org/10.1046/j.1469-0705.1995.05060424-2.x>.

16. McCullough, L. B., & Chervenak, F. A. (1994). *Ethics in obstetrics and gynecology.* New York: Oxford University Press, Inc. <https://doi.org/10.1046/j.1469-0705.1995.05060424-2.x>.

17. McCullough, L. B., & Chervenak, F. A. (1994). *Ethics in obstetrics and gynecology.* New York: Oxford University Press, Inc. <https://doi.org/10.1046/j.1469-0705.1995.05060424-2.x>.

18. McIntyre, A. (Fall, 2009). Doctrine of double effect. In E.N. Zalta (Ed.). *The Stanford Encyclopedia of Philosophy.* Retrieved from <http://plato.stanford.edu/archives/fall2009/entries/double-effect/>.

19. Wikipedia Contributors. (n.d.). *Institutional review board.* Wikipedia: The Free Encyclopedia. From <https://en.wikipedia.org/wiki/Institutional_review_board> Retrieved 16.10.18.

20. Miller, F. G., & Wertheimer, A. (2007). Facing up to paternalism in research ethics. *Hastings Center Report, 37*(3), 24−34. <https://doi.org/10.1353/hcr.2007.0044>.

21. Rothenberg, K. (1996). The Institute of Medicine's report on women and health research: Implications for IRBs and the research community. *IRB: A Review of Human Subjects Research, 18*(2), 1−3. <https://doi.org/10.2307/3563549>.

22. Weijer, C., Dickens, B., & Meslin, E. M. (1997). Bioethics for clinicians: 10. Research ethics. *Canadian Medical Association Journal, 156*(8), 1153−1157, citing Levine, R. J. (1997). *Ethics and regulation of clinical research.* New Haven, CT: Yale University Press.

23. Weijer, C., Dickens, B., & Meslin, E. M. (1997). Bioethics for clinicians: 10. Research ethics. *Canadian Medical Association Journal, 156*(8), 1153−1157, citing Levine, R. J. (1997). *Ethics and regulation of clinical research.* New Haven, CT: Yale University Press.

24. Kass, N. E., Taylor, H. A., & King, P. A. (1996). Harms of excluding pregnant women from clinical research: The case of HIV-infected pregnant women. *Journal of Law, Medicine & Ethics, 24*(1), 36−46. <https://doi.org/10.1111/j.1748-720X.1996.tb01831.x>.

25. Weijer, C., Dickens, B., & Meslin, E. M. (1997). Bioethics for clinicians: 10. Research ethics. *Canadian Medical Association Journal, 156*(8), 1153−1157, citing Levine, R. J. (1997). *Ethics and regulation of clinical research.* New Haven, CT: Yale University Press.

26. Lyerly, A. D., Little, M. O., & Faden, R. R. (2011). Reframing the framework: Toward fair inclusion of pregnant women as participants in research. *The American Journal of Bioethics, 11*(5), 50−52. <https://doi.org/10.1080/15265161.2011.560353>.

27. Eckenwiler, L. A., Ells, C., Feinholz, D., & Schoenfeld, T. (2008). Hopes for Helsinki: Reconsidering vulnerability. *Journal of Medical Ethics, 34*(10), 765−766. <https://doi.org/10.1136/jme.2007.023481>.

28. Eckenwiler, L. A., Ells, C., Feinholz, D., & Schoenfeld, T. (2008). Hopes for Helsinki: Reconsidering vulnerability. *Journal of Medical Ethics*, *34*(10), 765−766. <https://doi.org/10.1136/jme.2007.023481>.

29. Tong, R., & Williams, N. (2018, Summer). Feminist ethics. In E. Zalta (Ed.). *The Stanford Encyclopedia of Philosophy*. Retrieved from <https://plato.stanford.edu/archives/sum2018/entries/feminism-ethics/>.

30. Tong, R., & Williams, N. (2018, Summer). Feminist ethics. In E. Zalta (Ed.). *The Stanford Encyclopedia of Philosophy*. Retrieved from <https://plato.stanford.edu/archives/sum2018/entries/feminism-ethics/>.

31. Tong, R., & Williams, N. (2018, Summer). Feminist ethics. In E. Zalta (Ed.). *The Stanford Encyclopedia of Philosophy*. Retrieved from <https://plato.stanford.edu/archives/sum2018/entries/feminism-ethics/>.

32. Tong, R., & Williams, N. (2018, Summer). Feminist ethics. In E. Zalta (Ed.). *The Stanford Encyclopedia of Philosophy*. Retrieved from <https://plato.stanford.edu/archives/sum2018/entries/feminism-ethics/>.

33. Burton, B. K., & Dunn, C. P. (1996). Feminist ethics as moral grounding for stakeholder theory. *Business Ethics Quarterly*, *6*(2), 133−147.

34. Gilligan, C. (1987). Moral orientation and moral development. In E. F. Kittay, & D.T. Meyers (Eds.), (1989). *Women and moral theory*. Lanham, MD: Rowman & Littlefield Publishers, Inc.

35. McCullough, L. B., & Chervenak, F. A. (1994). *Ethics in obstetrics and gynecology*. New York: Oxford University Press, Inc. <https://doi.org/10.1046/j.1469-0705.1995.05060424-2.x>.

36. McCullough, L. B., & Chervenak, F. A. (1994). *Ethics in obstetrics and gynecology*. New York: Oxford University Press, Inc. <https://doi.org/10.1046/j.1469-0705.1995.05060424-2.x>.

37. Kukla, R. (2005). Conscientious autonomy: Displacing decisions in health care. *Hastings Center Report*, *35*(2), 34−44. <https://doi.org/10.1353/hcr.2005.0025>.

38. Kukla, R. (2007). How do patients know? *Hastings Center Report*, *37*(5), 27−35. <https://doi.org/10.1353/hcr.2007.0074>.

39. Noddings N. (1984). *Caring: A feminine approach to ethics and moral education*. Berkeley, CA: University of California Press.

40. Gilligan, C. (1987). Moral orientation and moral development. In E. Kittay, & D. Meyers (Eds.), *Women and moral theory*. Lanham, MD: Rowman & Littlefield Publishers, Inc.

41. Gilligan, C. (1982). *In a different voice: Psychological development and women's development*. Cambridge, MA: Harvard University Press in Feminist Ethics.

42. Jaggar, A. M. (1992). Feminist ethics. In L. Becker, & C. Becker (Eds). *Encyclopedia of ethics*. New York: Garland Press, cited in Tong, R., & Williams, N. (2018 Winter). Feminist ethics. In E.N. Zalta (Ed.). *The Stanford Encyclopedia of Philosophy*. <https://plato.stanford.edu/archives/win2018/entries/feminism-ethics>.

43. Fieser, J. (1996). Do businesses have moral obligations beyond what the law requires? *Journal of Business Ethics*, *15*, 457−468. <https://doi.org/10.1007/BF00380365>.

44. Freeman, R. E., & Gilbert, D.R. (1992). Business, ethics and society: A critical agenda. *Business & Society*, *31*(1), 9−17. <https://doi.org/10.1177/000765039203100102>.

45. Burton, B. K., & Dunn, C. P. (1996). Feminist ethics as moral grounding for stakeholder theory. *Business Ethics Quarterly*, *6*(2), 133−147.

46. Burton, B. K., & Dunn, C. P. (1996). Feminist ethics as moral grounding for stakeholder theory. *Business Ethics Quarterly*, *6*(2), 133−147.

47. Burton, B. K., & Dunn, C. P. (1996). Feminist ethics as moral grounding for stakeholder theory. *Business Ethics Quarterly*, *6*(2), 133−147.

48. Koski, E. G. (2005). Renegotiating the grand bargain: Balancing prices, profits, peoples, and principles. In M. A. Santoro, & T. M. Gorrie (Eds.), *Ethics and the pharmaceutical industry*. New York: Cambridge University Press.

49. Koski, E. G. (2005). Renegotiating the grand bargain: Balancing prices, profits, peoples, and principles. In M. A. Santoro, & T. M. Gorrie (Eds.), *Ethics and the pharmaceutical industry*. New York: Cambridge University Press.

50. Lupton, M. G. F., & Williams, D. J. (2004). The ethics of research in pregnant women: Is maternal consent sufficient? *British Journal of Obstetrics and Gynecology, 111*, 1307−1312. <https://doi.org/10.1111/j.1471-0528.2004.00342.x>.

51. Appelbaum, P. S., Roth, L. H., Lidz, C. W., Benson, P., & Winslade, W. (1987). False hopes and best data: Consent to research and the therapeutic misconception. *Hastings Center Report, 17*(2), 20−24.

52. Miller, F. G., & Wertheimer, A. (2007). Facing up to paternalism in research ethics. *Hastings Center Report, 37*(3), 24−34. <https://doi.org/10.1353/hcr.2007.0044>.

53. Pole, M., Einarson, A., Pairaudeau, N., Einarson, T., & Koren, G. (2000). Drug labeling and risk perceptions of teratogenicity: A survey of pregnant Canadian women and their health professionals. *Journal of Clinical Pharmacology, 40*, 573−577.

54. Koren, G., & Levichek, Z. (2002). The teratogenicity of drugs for nausea and vomiting of pregnancy: Perceived versus true risk. *American Journal of Obstetrics & Gynecology, 186* (5), S248−S252. <https://doi.org/10.1067/mob.2002.122601>.

55. Goldfarb, N. (2006). The two dimensions of subject vulnerability. *Journal of Clinical Research Best Practices, 2*(8), 1−3.

56. American College of Obstetricians and Gynecologists. (2015). ACOG Committee Opinion No. 646: Ethical considerations for including women as research participants. *Obstetrics & Gynecology, 126*, e100−e107. <https://doi.org/10.1097/AOG.0000000000001150>.

57. American College of Obstetricians and Gynecologists. (2015). ACOG Committee Opinion No. 646: Ethical considerations for including women as research participants. *Obstetrics & Gynecology, 126*, e100−e107. <https://doi.org/10.1097/AOG.0000000000001150>.

58. Harris, L. H. (2003). The status of pregnant women and fetuses in U.S. criminal law. *Journal of the American Medical Association, 289*(13), 1697−1699. <https://doi.org/10.1001/jama.289.13.1697>.

59. Kolder, V. E., Gallagher, J., Parsons, M. T. (1987). Court-ordered obstetrical interventions. *New England Journal of Medicine, 316*, 1192−1196. <https://doi.org/10.1056/NEJM198705073161905>.

60. American College of Obstetricians and Gynecologists. (2016). ACOG Committee Opinion No. 664: Refusal of medically recommended treatment during pregnancy. *Obstetrics and Gynecology, 127*, e175−e182. <https://doi.org/10.1097/AOG.0000000000001485>.

Part II

Quantitative and qualitative discoveries

Chapter 4

A measure of exclusion

Clinical trials have traditionally excluded pregnant women to protect their fetuses from exposure to experimental medications with unknown safety risks. But this practice has contributed to a lack of information about how to treat medical conditions that can complicate pregnancies. This does not mean that all studies should include pregnant women. Reasons for their exclusion should be considered prior to study initiation. For example, it would be reasonable to exclude pregnant women from drug studies for which therapeutic benefit is undetermined.

Including pregnant women in clinical trials would result in better information about the safety and efficacy of treatment options during pregnancy and this has been suggested by experts who are recommending their inclusion.[1,2,3,4] But there is little information from the US pharmaceutical industry about the status of pregnant women's participation in clinical research studies. How many drug trials actually include pregnant women? To better understand the US pharmaceutical industry's practices and perspectives about the inclusion of pregnant women in clinical research, I did some researching myself.

Confirming and quantifying the exclusion of pregnant women from clinical studies would provide a view into current practice. In addition, if the US Food and Drug Administration (FDA) guidance is implemented by the pharmaceutical industry, the measure can serve as a baseline for the future evaluation of its impact.

The US clinical trial system

Fig. 4.1 depicts the clinical trial system in the United States. Before drugs in development even reach Phase 1 studies, preclinical studies in animals are performed. (Trials testing potential new medicines with animal subjects are called preclinical trials; trials testing potential new medicines with human subjects are called clinical trials.) An important part of this preclinical testing is the evaluation of a drug candidate's potential for toxicity and teratogenic effects on the developing animal fetus. After the completion of all the preclinical studies, FDA weighs the pregnancy risk information to assess the

Pregnancy and the Pharmaceutical Industry. DOI: https://doi.org/10.1016/B978-0-12-818550-6.00004-0

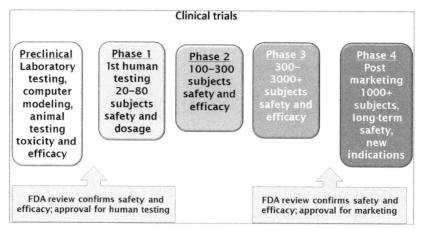

FIGURE 4.1 US clinical trial progression.[5]

safety of the new drug entity on the animal fetus prior to giving approval for the drug to be tested in humans.

Potential drug candidates are tested for the first time in humans in Phase 1 studies. Typically fewer than 100 study subjects participate. The purpose is to evaluate the human response to the drug, its dosage and safety in a small, very well-controlled environment. Based on the results, about 70% of these drugs move on to the next phase of study; the others are withdrawn.

Up to several hundred people who have the disease the drug will potentially treat participate in Phase 2 studies. This phase of testing can last from several months up to 2 years and looks at the efficacy of the drug against the condition and its side effects. About 33% of drug candidates will pass this testing.

Phase 3 trials are large, enrolling from 300 to over 3000 volunteers, and can last from 1 to 4 years. They evaluate the efficacy and side effects of the drug candidate in a large number of people with the disease. About 25% of these drugs will ultimately be approved by FDA.

After the drug is approved by FDA and is on the market additional studies may be conducted. Several thousand people who need treatment can be enrolled in these "Phase 4" or "postmarketing" clinical trials. These studies provide additional information about the safety and efficacy of the drug in actual usage by a more diverse population than were studied in earlier clinical trials. In addition to safety and efficacy, these studies may be designed to evaluate the drug's effectiveness in certain populations, potential new uses for the drug, or the potential for drug interactions. Phase 4 studies can continue for many years after the drug is on the market. Some are conducted voluntarily by the drug company, others are mandated by FDA for further evaluation of safety issues.

Because these drugs have been studied in clinical trials, approved by FDA, and are on the market, more safety and efficacy information is available and the benefit–risk relationship is better established. Phase 4 studies, therefore, are some of the most appropriate studies in which pregnant women could participate. The exclusion of pregnant women from Phase 4 studies could serve as a proxy for the practice of exclusion of pregnant women from all phases of clinical trials. Looking at their exclusion from Phase 4 trials could provide a conservative estimation of their exclusion from all phases of drug trials. In other words, if they are not allowed to participate in Phase 4 drug studies, it is even more unlikely that they are being asked to participate in Phase 1, 2, and 3 studies.

Are pregnant women excluded from US clinical trials?

As discussed in Chapter 1, Drug testing and pregnant women: background and significance, there is no restriction upon pharmaceutical companies from enrolling pregnant women in drug studies providing that the parameters of The Common Rule are met. As the responsible entity for the cost, conduct, and outcome of clinical trials, pharmaceutical companies are key decision-makers that determine whether pregnant women will be included in or excluded from enrollment in each study. To get a measure of their actual exclusion, my question was, "What proportion of open US pharmaceutical industry-sponsored Phase IV clinical trials currently enrolling women include pregnancy in their exclusion criteria?"

I found it was possible to quantify the frequency with which pregnant women are excluded from current Phase 4 clinical trials evaluating treatments for diseases or conditions that could affect a pregnancy by reviewing the inclusion/exclusion criteria of all US based, open Phase 4 interventional drug studies with adult female participants sponsored by industry currently posted on ClinicalTrials.gov.

My data source was the inclusion/exclusion criteria of all US-based, industry-sponsored open Phase 4 interventional drug studies enrolling women of childbearing potential posted on the www.ClinicalTrials.gov website between October 1, 2011 and January 31, 2012.

ClinicalTrials.gov provides public access to the list of federally and privately supported clinical trials currently being conducted to investigate experimental treatment for a wide range of diseases and conditions. The website was developed to provide public information about current clinical trials so that individuals with serious diseases and conditions might access experimental treatments and volunteer to participate in the studies. It also provides a resource to access the basic results of completed clinical trials.

According to its website, over 287,236 research studies from all 50 states and 204 countries have been registered on ClinicalTrials.gov.[6] It gets over 61,000 visitors/day and is updated nightly.[7] Studies listed in the database were conducted in all 50 states and in 174 countries. The NIH, through its National Library of Medicine, has developed this site in collaboration with the FDA, as a result of the FDA Modernization Act, which was passed into law in November 1997.

I limited the dataset to the open (i.e., currently enrolling) US-based Phase 4 (postmarketing) interventional studies (i.e., studies that provide participants with new or known or no therapies to determine whether the treatments are safe and effective in a controlled environment as opposed to observational studies that monitor a group of participants who have selected their own treatment to determine whether the treatment is safe and effective in their own environment) that included adult (i.e., age 18−65) female participants and were sponsored by a pharmaceutical company. To be included, the study must have been evaluating treatment of conditions that may be experienced by, but are not limited to, pregnant women and they must not have involved the use of a medication that is in the FDA pregnancy categories D or X—those thought to be potentially teratogenic.

During the process of a medication's approval, and prior to marketing, FDA weighs the pregnancy risk information attained in animal studies and clinical trials and, up until December 2014, applied a pregnancy risk category (see Table 4.1) to the products' label. At that time, FDA implemented a phase-out of pregnancy categories and mandated a narrative description of the drug's effects on animal offspring and in any (usually inadvertent) pregnancies that may have occurred in clinical trials.[8] Later pregnancy exposure experience would be added to the information in the label on an ongoing basis. Pregnancy categories are to have been removed from all drug labels

TABLE 4.1 Former US Food and Drug Administration pregnancy risk categories.[9]

Risk	Category
A	Adequate and well-controlled studies in pregnant women have failed to demonstrate a risk to the fetus in the first trimester of pregnancy
B	Animal reproduction studies have failed to demonstrate a risk to the fetus and there are no adequate and well-controlled studies in pregnant women **or** animal reproduction studies have shown adverse effects, but well-controlled studies in pregnant women have shown no adverse effects to the fetus
C	Animal reproduction studies have shown an adverse effect on the fetus, **or** there are no animal reproduction studies and no well-controlled studies in humans
D	Positive evidence of fetal risk, but benefits may outweigh risks
X	Positive evidence of fetal risk, and risks clearly outweigh any possible benefit

by 2020. At the time of this evaluation, Pregnancy Risk Categories were still in use for all drugs being studied in Phase 4 trials.

Data acquired from ClinicalTrials.gov included: Study ID#, name of study, drugs, sponsor (pharmaceutical company name), condition, estimated enrollment, inclusion criteria, exclusion criteria, contact information, and study description. The FDA pregnancy risk category was added by checking each medication's Physician's Package Insert on www.PDR.net that is provided by the pharmaceutical company and approved by FDA.

Clinical trials were excluded from the analysis if the age (e.g., postmenopausal), condition (e.g., amenorrhea), or drugs being tested (e.g., pregnancy risk category D or X) would preclude the enrollment of pregnant women.

Phase 1, 2, and 3 trials were excluded because the safety data of the unapproved drugs being evaluated are not available in the public domain. Therefore, I would have been unable to evaluate if it would be appropriate for pregnant women to be included in such studies.

However, this review could not have addressed de facto exclusion, that is, the "inadvertent" failure to recruit pregnant women (as opposed to exclusion de jure, i.e., the explicit exclusion of pregnant women in the protocol).[10] In other words, just because a study did not exclude pregnant women in the protocol, did not mean that pregnant women would be enrolled. If there was no mention of pregnancy in the inclusion or exclusion criteria of a study, a study coordinator was contacted to confirm that pregnant women could be enrolled.

The measure of their exclusion

Application of the search criteria above retrieved a total of 558 studies from the ClinicalTrials.gov website. ClinicalTrials.gov is an open website, with clinical trials being added or removed on a continuous basis as new trials are initiated and completed studies are removed. Therefore, the number of trials retrieved on any given day may vary. This point-in-time examination was limited to those studies that were contained in the database as of October 1, 2011 and up to and including January 31, 2012.

Of the 558 studies, 103 were excluded from further analysis: 5 studies limited enrollment to pregnant women and 98 were found to appropriately exclude pregnant women based on the following justifications:

- $n = 74$—At least one drug in study was in FDA pregnancy category D or X.
- $n = 17$—Age criteria excluded childbearing potential.
  - 10 studies limited inclusion to men and women age ≥ 60.
  - 6 studies limited inclusion to men and women age ≥ 50.
  - 1 study limited inclusion to men and postmenopausal women age ≥ 45.
- $n = 4$—Study topic was menopause.
- $n = 2$—Study topic was contraception.
- $n = 1$—Study topic was lactation.

Of the remaining 455 clinical trials that could potentially enroll women of childbearing potential, 301 specifically excluded pregnant women, 1 specifically did not exclude pregnant women, and 153 made no mention of pregnancy in the inclusion/exclusion criteria posted on ClinicalTrials.gov.

One might expect that, if pregnancy was not mentioned in the exclusion criteria, then pregnant women could be included. In order to test that assumption, all 153 of these studies' coordinators were contacted by phone, by email, or both. Eighty-eight did not respond to the request for information. Of the 65 studies for which I obtained clarification about whether or not pregnant women could be included, 47 (72%) actually excluded pregnant women and 18 (28%) did not exclude them from enrollment in the study (Fig. 4.2).

The overall results of the analysis were that, of the 367 clinical trials for which specific inclusion/exclusion criteria were obtained, 348 (95%) excluded pregnant women and 19 (5%) did not. These 367 Phase 4 studies in which pregnant women could appropriately participate confirmed that excluding pregnant women from clinical studies in which they could safely participate is common practice.[11]

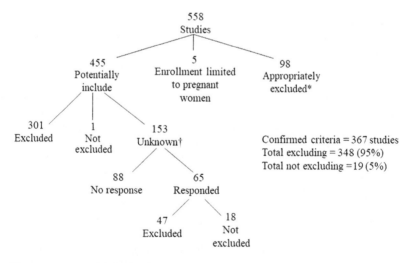

Number of studies excluding and not excluding pregnant women

*Pregnant women were excluded because the drug was in FDA category D or X, or the age or topic (menopause, contraception, lactation) prohibited pregnancy.
†Study coordinators were contacted if enrollment criteria did not address pregnancy.

FIGURE 4.2 Enrollment of pregnant women in Phase 4 clinical studies available for review on ClinicalTrials.gov (October 1, 2011–January 31, 2012).

Discussion

A range of medical conditions that could occur in pregnant women comprised the Phase 4 studies evaluated. These included epilepsy, depression, fungal infection, arthritis, heart failure, peripheral artery disease, Von Willebrand disease, aneurysm, human immunodeficiency virus, etc. All of the conditions could occur in pregnant women, though treatment of some, like knee replacement, might be safely postponed until the completion of the pregnancy.

The topics of the five studies that were limited to women who were pregnant or intending to become pregnant were all pregnancy-related conditions. There were no studies identified in this review that were specifically designed to evaluate the treatment of nonobstetric illness during pregnancy.

Comparing the studies that were open to the enrollment of pregnant women to those that excluded pregnant women found more commonalities than differences. In approximately 80% of the studies in both groups, the purpose of the trial was to evaluate treatment as opposed to prevention, diagnostic, or supportive care. Masking (or blinding) was also similar. The proportion of studies that were blinded (subjects and/or researchers do not know what treatment the subject is receiving) as compared to those that were "open-label" (subjects and researchers are aware of the treatment that the subject is receiving) was also similar—42% of each.

The only variable noted to differ between the two groups was allocation—that is, whether the studies were randomized trials (in which subjects are randomly assigned to different treatment groups) or nonrandomized trials (subjects may choose, or are in previously chosen treatment groups). In the 348 studies that excluded pregnant women and reported allocation on ClinicalTrials.gov, 66% were randomized and 34% were not. In the small number of studies ($n = 19$) that reported allocation and did not exclude pregnant women, 53% of the trials' protocols randomized subjects into different treatment categories and 47% were nonrandomized. This suggests that researchers who are open to including pregnant women in their studies may be open to including them in both randomized and open-label studies.

By contacting the study coordinators for the trials that did not address pregnancy in the inclusion/exclusion criteria to clarify the protocol, I determined that the inclusion/exclusion criteria posted on ClinicalTrials.gov was often inaccurate in regards to pregnant women—an interesting finding in and of itself. Seventy-two percent of studies that did not list pregnant women in their exclusion criteria posted on ClinicalTrials.gov did, in fact, exclude them. This may infer that the practice of excluding pregnant women is so commonplace that it is simply assumed to be true and, therefore, did not need to be explicitly stated. Comments such as this one supported the assumption of exclusion: "We don't have to list every exclusion on ClinicalTrials.gov. No glaucoma treatments have been approved for use in pregnant women. So this [not to include pregnant women in the exclusion criteria] was a no-brainer."

Other studies may not have specified pregnancy in the criteria because the condition was assumed to have a low prevalence in the treatment group, i.e., certain age groups (adolescents, women over 40) or medical conditions (cardiomyopathy, dialysis). For example, one emailed response stated, "There is no specific pregnancy criteria in this study. The drug is mostly used in dialysis pts, who are generally unable (with rare exception) to become pregnant anyway. NOTE: Just because not excluded doesn't mean they will be included." This quote addresses the issue of de facto exclusion: pregnancy may not be an exclusion criterion, but that doesn't mean pregnant women will be enrolled. This is reflected in the finding that only one of the nineteen studies (5%) that included pregnant women specifically stated so in the inclusion criteria. So the de facto exclusion may work both ways— whether a pregnant woman can be enrolled in a trial is frequently not explicitly stated in the inclusion/exclusion criteria and if she is not mentioned in either category, she still may not be eligible, and if eligible, she still may not be included.

What about the omission of the 88 studies that did not list pregnancy in the exclusion criteria but did not respond to attempts at clarification? What if they may have included them? I applied the same ratio of exclusion to inclusion that I found for the 65 with confirmed criteria (72% excluded vs 28% not excluded) to these 88 studies. That calculation added an additional 63 studies to the number that excluded women and 25 to the number that did not. This resulted in an estimated total of 411 studies that excluded pregnant women and 44 that did not. In this scenario, of the 455 Phase 4 studies in which pregnant women could potentially participate, 90% excluded and 10% did not exclude pregnant women from enrollment. Whether the actual proportion is the 90% I could estimate from this calculation or the 95% that I confirmed by the study monitors—or somewhere in between—does not impact the results: the exclusion of pregnant women from clinical research is clearly the norm.

Two studies directly addressed the need for and the appropriateness of including pregnant women. One recognized that pregnancy would rarely occur in women with heart failure but had other studies to address that issue: "There is not a specific exclusion for pregnant women in [this] trial. However, this would be an extremely rare occurrence ... We have other trials for pregnant women with cardiomyopathy." The other was the study that explicitly included pregnant women in the inclusion criteria: "[Drug] carries a Category B pregnancy risk factor. Since this is a minimal pregnancy risk category, no special precautions will be taken to determine that the patient is not pregnant." These are two examples of an appropriate approach to the enrollment of pregnant women in clinical trials. The first acknowledges that cardiomyopathy may occur during pregnancy and has designed studies for the study of the treatment in pregnant women. The enrollment criteria refer them to the other studies. The second study explicitly conveys that there is no rational reason for their exclusion so pregnant women would be eligible to enroll. Here, the relatively new pregnancy labeling rule would be helpful. The drug label would include a description of the completed animal and human studies and would be available for any researcher or potential participant to review and assess the risk, rather than a vague "category B."

It would be difficult to include an exhaustive list of conditions that would preclude any subject from being eligible for enrollment on the ClinicalTrials. gov website. Some study protocols addressed this issue by including statements such as "Additional inclusion (exclusion) criteria may apply," "any subject who at the discretion of the Investigator is not suitable for inclusion in the study," or "has any other clinically important abnormalities such that risk to patient of participation outweighs the potential benefit of therapy as determined by the investigator." Such statements leave the enrollment decision up to the individual principal investigator (PI). Contacting the study coordinators or sponsors to confirm the number of studies that actually excluded pregnant women resulted in a more accurate assessment. A 95% exclusion rate confirms a very common practice. But of the 5% that could enroll them, I don't know how many actually did.

A comparison of the categories of the treatments under study was also made. There were six major categories: drugs ($n = 244$), devices ($n = 77$), procedures ($n = 25$), biological products ($n = 16$), and diet ($n = 5$). Ninety-eight percent of drug studies excluded pregnant women, 92% of studies on procedures excluded pregnant women, 90% of device trials excluded pregnant women, and 81% of biological product studies excluded them. It was not surprising to find that drug trials excluded pregnant women at a rate that is higher than that for procedures, devices, and biologics. The precaution against the testing of drugs on pregnant women is very well-entrenched. However, the overall magnitude of the exclusion (98%) in these Phase 4 drug studies in which the pregnancy risk categories were defined, is higher than anticipated. Of interest was the finding that 19% of the studies on biological agents did not exclude them, the highest of any category. This may be reflected in our increased acceptance of vaccines during pregnancy.

Two vaccines are currently recommended for administration during pregnancy: influenza vaccine to protect against the flu (can be given at any time during pregnancy) and Tdap vaccine to protect against tetanus, diphtheria, and pertussis (after 20 weeks gestation). Other vaccines including those against hepatitis, meningococcal disease and pneumococcal disease, may be advisable depending on individual circumstances. Under development are vaccines for use during the third trimester of pregnancy that provide antibodies for the newborn, including one for respiratory syncytial virus (RSV) and one for group B streptococcus (GBS).[12] These two diseases are responsible for a high percentage of the neonatal and infant deaths that occur annually worldwide.

At the time of this publication, the RSV vaccine candidate is being tested with pregnant women in their third trimester. The researchers planned to enroll over 8000 participants "to determine the efficacy of maternal immunization with the RSV F vaccine against symptomatic RSV lower respiratory tract infection with hypoxemia through the first 90 days of life in infants."[13] These women will be followed to 6-month postdelivery and their infants will be monitored for approximately 1 year.

The GBS vaccine candidate is also currently being tested with pregnant women. The focus of one current trial is to "Evaluate the safety and immunogenicity of the trivalent group B streptococcus vaccine in healthy pregnant women [and] evaluate the levels of GBS serotype-specific antibodies in infants, placental transfer from the pregnant women to the infant and levels of antibodies in the breast milk."[14]

The acceptance of maternal immunization to improve protection of newborns from infectious disease may have a positive impact on the acceptance of immunizations in general during pregnancy—and potentially on the acceptance of clinical studies including pregnant women as well.

Of course, all clinical studies should not include pregnant women. For example, studies for conditions for which treatment could be postponed might recommend deferral of enrollment until the conclusion of the pregnancy. I did not attempt to exclude studies for conditions in which treatment can be deferred because, in many instances, that conclusion could be determined only by the patient and her health care provider. Severity of the condition, its effect on the patient's quality of life, and alternative palliative treatment options are factors that may need to be considered in that determination.

Application of the findings

Phase 4 clinical trials are conducted using drugs, devices, and other treatments that have received regulatory approval and are currently marketed in the United States. Phase 4 studies are conducted for various purposes including, but not limited to, evaluating safety issues, long-term effects, cost-effectiveness, and use in populations or for conditions other than those for which the drug was initially approved. Because these drugs and devices have been studied in clinical trials, were approved by FDA, and are on the market, safety, efficacy, and the benefit−risk relationship are better established. These studies, therefore, are some of the most appropriate studies in which pregnant women could participate.

Women with medically compromised pregnancies and their health care providers can be frustrated by a lack of clinical data available to inform treatment decisions. Subject matter experts state that including pregnant women in clinical trials would result in better information about treatment options during pregnancy. The medical literature provided scant documentation or estimate of the proportion of clinical trials that exclude pregnant women.

Awareness of the extent of their exclusion from industry-sponsored trials will provide a baseline against which to measure future progress. I established that approximately five percent of current industry-sponsored trials include pregnant women among the populations eligible for enrollment. However, the overwhelming proportion—95%—of studies that exclude pregnant women suggests that, even when it may be appropriate to include them, the practice of excluding them is very well-established.

Notes

1. Feibus, K., & Goldkind, S. F. (2011). Pregnant women and clinical trials: Scientific, regulatory, and ethical considerations. In *Oral presentation at the pregnancy and prescription medication use symposium.* May 17, 2011, Silver Springs, MD.
2. Institute of Medicine, Mastroianni, A. C., Faden, R., & Federman, D. (Eds.). (1994). *Women and health research: Ethical and legal issues of including women in clinical studies.* Washington, DC: National Academy Press.
3. *Council for International Organizations of Medical Sciences. CIOMS international ethical guidelines for health-related research involving humans.* (2016). Retrieved from <https://cioms.ch/wp-content/uploads/2017/01/WEB-CIOMS-EthicalGuidelines.pdf>.
4. American College of Obstetricians and Gynecologists. (2015). ACOG Committee Opinion No. 646: Ethical considerations for including women as research participants. *Obstetrics & Gynecology, 126,* e100−e107. https://doi.org/10.1097/AOG.0000000000001150.
5. *U.S. Food and Drug Administration. The drug development process, step 3: Clinical research.* (2018). <https://www.fda.gov/forpatients/approvals/drugs/ucm405622.htm> Retrieved 17.10.18.
6. *U.S. National Library of Medicine. ClinicalTrials.gov.* (n.d.). <https://clinicaltrials.gov/> Retrieved 17.10.18.
7. William, R. J. (2015). *Get to know ClinicalTrials.gov. [Powerpoint® presentation].* Retrieved from <https://www.fda.gov/downloads/ForConsumers/ByAudience/MinorityHealth/UCM465711.pdf>.
8. U.S. Food and Drug Administration. (2014). Content and format of labeling for human prescription drug and biological products: Requirements for pregnancy and lactation labeling. *Federal Register, 79*(233), 72064−72103.
9. *Drugs.com. FDA pregnancy risk categories prior to 2015.* (2018). <https://www.drugs.com/pregnancy-categories.html> Retrieved 17.10.18.
10. Merton, V. (1993). The exclusion of pregnant, pregnable, and once-pregnable people (a.k.a. women) from biomedical research. *American Journal of Law & Medicine, 19*(4), 369−451.
11. Shields, K. E., & Lyerly, A. D. (2013). Exclusion of pregnant women from industry-sponsored clinical trials. *Obstetrics & Gynecology, 122*(5), 1077−1081. https://doi.org/10.1097/AOG.0b013e3182a9ca67.
12. Vojtek, I., Dieussart, I., Doherty, T. M., Franck, V., Hanssens, L., Miller, J., et al. (2018). Maternal immunization: Where are we now and how to move forward? *Annals of Medicine, 50*(3), 193−208. https://doi.org/10.1080/07853890.2017.1421320.
13. *Novavax. A study to determine the safety and efficacy of the RSV F vaccine to protect infants via maternal immunization.* (2015). <https://clinicaltrials.gov/ct2/show/NCT02624947?term = Novavax> Retrieved 17.10.18.
14. *GlaxoSmithKlein. Safety and immunogenicity of a trivalent group B streptococcus vaccine in healthy pregnant women.* (2014). <https://clinicaltrials.gov/ct2/show/NCT02046148> Retrieved 17.10.18.

Chapter 5

Perspectives from the industry: on exclusion

In order to isolate the perspectives of the people who work in the pharmaceutical industry and in other organizations that are involved in clinical trials, I used stakeholder interviews to identify their rationales for excluding pregnant women from clinical research and to discover if they thought there were opportunities for their inclusion. The interviews were conducted in 2012, while a guidance on the inclusion of pregnant women in clinical trials was on the FDA docket for publication,[1] but before it was released to the public in 2018.

Groundwork

To obtain stakeholders information, 16 interviews were conducted with stakeholders (subject matter experts) in the US pharmaceutical industry and related organizations including:

- Staff from clinical development, safety, regulatory, and epidemiology departments in US pharmaceutical companies ($n = 5$), biotech firms ($n = 2$), and a contract research organization ($n = 1$)
- Pharmaceutical Research and Manufacturers of America (PhRMA), (the industry professional association) member ($n = 1$)
- Legal counsel from pharmaceutical companies ($n = 2$) and PhRMA ($n = 1$)
- Independent Institutional Review Board (IRB) members who are contracted by the industry to review clinical research protocols ($n = 3$)
- Food and Drug Administration employee ($n = 1$)

Note: BIO, the biotechnology industry association, was invited but declined to have an association member participate in the interviews.

A total of 12 unique companies and organizations were included in the discussions. Questionnaires were sent to the stakeholders prior to the

Pregnancy and the Pharmaceutical Industry. DOI: https://doi.org/10.1016/B978-0-12-818550-6.00005-2

interviews. It is important to note that the participants were speaking on behalf of themselves and were not representing the position of their companies or organizations. I provided assurance that their names would be kept confidential and, with their permission, 14 of the 16 interviews were recorded (2 lawyers declined to be recorded).

Stakeholders on exclusion

The primary purpose of the interviews was to identify from stakeholders in the pharmaceutical industry and related organizations their perceived rationales for the exclusion of pregnant women from clinical research and to discover opportunities for their broader inclusion. The focus of this chapter is stakeholders' perceptions of exclusion. Chapter 6, Perspectives from the industry: on inclusion, will focus on inclusion.

One of the key missing pieces of information identified by the 2009 Second Wave Consortium workshop, for which this project sought clarification was, "What is the strength of the pharmaceutical industry's role in affecting the outcome of the debate on the inclusion of pregnant women in clinical trials?"

Question 1. Control

Who has the most control over whether pregnant women are included in clinical trials—the pharmaceutical company (the Sponsor), the IRB, or FDA?

All respondents acknowledged that all three stakeholders have influence and share responsibility for the decision to include or exclude pregnant women from clinical trials.

The Sponsor has to be willing to conduct the research, the FDA has to be willing to sanction it, and the IRB has to approve it. It's like a pyramid or a triangle. We all have to have buy-in. The Sponsors ... write the protocols and provide the inclusion and exclusion criteria. IRB has absolute power to change it. FDA has absolute power to put a clinical hold on it. Any one of them could stop the study from moving forward.   —IRB member

I would say that any of the three probably have a veto in the sense that pregnant women are not going to be enrolled in a trial unless all three of those groups agree.   —Pharma attorney

However, most also agreed that, although all three stakeholders have the ability to influence the final decision, ultimately it is the Sponsor who has the most control. Reasons cited by the participants for the pharmaceutical company's dominance included that the Sponsor:

- Decides if they want to sponsor the trial or not.
- Is the owner of the protocol and makes the initial eligibility criteria decisions.
- Is the driver of what is to be accomplished by the study.
- Knows the science behind the drug in development.
- Is the responsible party that would bear the burden of any drug exposure–perceived injury.

> ...an IRB can be comfortable or uncomfortable, the FDA can be comfortable or uncomfortable, but it seems that the people running the clinical studies, sponsoring the clinical trials in pharma, are the ones who will bear the immediate responsibility if something is perceived as going wrong—that may or may not be related to the drug exposure.   —Pharma member

Also, the stakeholders felt that the consideration of the inclusion of pregnant women would only arise if the Sponsor suggested their inclusion. The FDA and the IRB would then respond to the proposal. They felt that, at this point in time, it would be unlikely for a Sponsor to exclude pregnant women and have FDA or the IRB recommend that they be included.

> The FDA and the IRB can't force the company to include pregnant women in the clinical trials because ultimately it's a question of liability. So they're not the ones who are accepting the liability risk.   —Pharma physician

Not everyone agreed, however, and a few perceived the IRB as having the most control. A pharmaceutical company lawyer said,

> ...ultimately the IRB is the one that primarily has [the decision on] the inclusion of pregnant women; it is fundamentally an ethical issue that the IRB ... has ownership of. But I think all three really share responsibility.
>   —Pharma attorney

> The company can write the protocol, they can present it to the FDA, but the last word of approval really comes from the IRB.   —IRB member

Another suggested the IRB and FDA together held the most power.

> I think it's the IRB and the FDA [that] have ... more control because even if the sponsor wanted to set up an arm or a sub-arm to look at pregnant women, there's an absolute need for FDA approval of that part of the protocol and then getting the IRB to approve it.   —PhRMA member

Overall, FDA was perceived to have the least influence on the decision to include pregnant women in clinical trials.

As far as the FDA, my impression is that it may not be common for them to comment but [they] would defer to the IRB since there is no regulatory prohibition. They would take a secondary role. —Biotech attorney

IRB review was noted to have some limitations. A company employee noted that IRBs only see one protocol at a time; they do not see the other protocols in development for the same product. They wouldn't know if a company was designing other studies that included pregnant women.

...one of the issues we work on at PhRMA, which still has a way to go, is for multicenter trials to get greater acceptance of using a central IRB simply because of the variance of IRBs in terms of accepting a trial in a number of areas.
—PhRMA member

Of note is the identification of a fourth important stakeholder: the research institution.

The IRB can review and approve the study and ... feel it's perfectly safe and all the risks have been minimized, and there is terrific benefit, but ... if the hospital's not okay with the conduct of the study, it will never happen. So you need all four points today − the Sponsor, you need the IRB, you need the FDA, and you need the facility, too. —IRB member

Some participants volunteered their perceptions of how the other stakeholders would respond to the issue. Of the three major stakeholders, IRBs were felt to be the most cautious, followed by the regulatory agencies, followed by the Sponsors.

The biggest contemporary challenge would be ethics review committees, IRBs. I think in part it's because their focus is rather narrow and they are diverse bodies. I don't think they're always necessarily in step with contemporary thinking in clinical trials. Often they're very individual patient-focused and often even lose sight of the wider patient population focus. I still think that they're your biggest problem or controlling factor. —Pharma physician

Once we see the word pregnancy, we step back and take a really close look at this. There's just so many things that could go wrong and you don't want to have not considered all of the risk factors. —IRB member

A pharma physician thought the FDA would be inclined to approve the inclusion of pregnant women if you could specify why you felt the study would be of benefit to the patient population, that is, if there were few treatment options or had little data to inform dosing, and there was no known significant safety risk.

I am of the opinion that the FDA would be fine with that study. It's just been my experience that pharmaceutical companies, if you say pregnancy, everyone just sort of turns off, they switch off. It's like we don't really deal with that.

—Pharma physician

Key findings for Question 1

Among the pharmaceutical industry researchers and lawyers, IRBs, PhRMA stakeholders, and FDA, the consensus was that the pharmaceutical industry has a dominant role in influencing the outcome of this debate.

1. Four stakeholders were identified as having the power to veto a clinical trial: the trial's sponsor, the FDA, the IRB, and the institution at which the trial is to take place.
2. The Sponsor was perceived to have the most control over whether or not pregnant women were included as study subjects. Without proposing their inclusion, it was unlikely that the FDA or IRB would suggest it. In addition, it was felt that because the company would have the highest risk for liability, they had the right to be the decision-makers.
3. FDA was found to have the least influence as there is no regulatory statute that requires the exclusion or inclusion of pregnant women from clinical trials (they may be included under certain circumstances).
4. IRBs were felt to be a potential barrier to inclusion based on their cautious nature, their patient-centric focus, and the variability of decisions from one IRB to the next.

Question 2. Policy and practice

Does your organization have a policy about whether or not to include pregnant women in clinical studies? What are the current practices?

The practice is to exclude pregnant women. It's fairly entrenched.

—Biotech physician

The following quotes, from the pharma and biotech companies and lawyers in those companies, describe the status quo:

- *I can't name one company that has done clinical trials with pregnant women.*
- *Current practice is to exclude women who are known to be pregnant.*
- *The practice is to exclude pregnant women. It's fairly entrenched.*
- *I know that it is, in fact, our practice to exclude pregnant women except in therapies that are designed to assist pregnancy.*
- *All our trials exclude pregnant women.*

- *I'm not aware of an actual policy saying that they shouldn't be in but they just never are.*
- *I think in this day and age they'd be apprehensive about putting it in writing but I think today it's a widespread assumption and convention that you just don't do that.*
- *I am not aware of any studies by any Sponsor that would include pregnant women.*

Yet, there were a couple of pharma company employees that had experience with clinical studies that included pregnant women. One physician stated that companies "don't want something bad to happen to a pregnant woman or her fetus or her newborn relative to their drug. So they are cautious more than anything." Her experience involved products to treat HIV and to prevent vertical transmission to a fetus. Prior to the initiation of the company's studies, information on their drug's use in pregnant women was available from pregnancy registries and from NIH trials. With that safety information on hand, she was able to design additional studies for pregnant women. She stated,

> the problem with industry is, the people who are going to make the decisions, are they knowledgeable enough about the drugs, the safety profile of the drug, of the data, and the therapeutic area in need? If they understand all of those things, I think the discussions go a little bit faster and a little bit easier. But if they don't understand, if they don't have that knowledge base, then you have a lot of hurdles to overcome in industry.

Another company employee described the oversight boards and safety monitoring boards that the clinical trial protocols must go through prior to study initiation. "A study for pregnant women would have such oversight as well," he said. Even now, if a woman becomes pregnant on a trial and her PI feels that it would be in her best interest to continue on the study drug, the PI can request a protocol exclusion to keep her enrolled on a compassionate use basis. "Generally," he said, "the patient is immediately withdrawn but there is the opportunity to challenge that."

There is no PhRMA organization statement on the inclusion of pregnant women in clinical trials. The organization person I spoke with stated that this issue has never come up in conversations with the FDA Office of New Drugs. He was unaware of the guidance in review at the agency on this topic before we spoke. He added, "quite frankly, most of what PhRMA does tends to be reactive rather than proactive. So that when FDA will issue a draft guidance for comment, PhRMA would then convene or develop a group … to prepare comments for it."

The IRBs, however, had more detailed responses to this question. Each IRB interviewee stated that they do have a policy on pregnant women in clinical trials—it is their policy on "vulnerable populations."

[The policy] talks about subpart B of subpart 45, it talks about various con-
cerns or considerations when including pregnant women. Generally, what it
outlines is that pregnant women may be included in a clinical trial — again, if
it's in compliance with the regulations, and also if the board thinks that it's
essentially ethically acceptable.

We do have a policy, because you can't leave them out of everything. Because
if you leave them out of everything, you'll never learn anything, and they have
the right to be studied just like anybody else does. You need information on
them like for everybody else. You can't exclude them because that would be
unfair. But you just have to be very careful when you do include them.

However, one of the IRB members considered minimization of risk to come before the equitable selection of subjects, stating that, "even before the Belmont principles really, beneficence and non-maleficence rise to the top of the list and autonomy is lower down." Another stated that, while their policy is somewhat general at this time, "it will be more elaborate if the guidance came out."

Additional IRB members commented on the draft FDA guidance:

I think that once they have laid out what they would consider clear-cut guide-
lines for the enrollment of pregnant women in research, I think that more IRBs
would be more willing to take a look. The FDA guidance will be helpful.

I think that as soon as there's guidance put out there, and one IRB reads it
and feels that they can actually provide proper oversight, that women will be
able to be enrolled in clinical trials even though they're pregnant.

The FDA interviewee, a former pharma company employee, stated:

Unless the company is dealing with some very specific disease area, I don't
believe companies will go there ... A mark of success to me, will be the com-
pany that has in its research plan for a chronic drug for depression, a plan to
enroll pregnant women post-marketing. But at this point it's going to be
restricted to those specialty areas where it is clear that the studies haven't
been done in the past and they'll be expected to do them. Like some of the
tropical diseases, the developing world concerns, malaria comes to mind.

Key findings for Question 2

The consensus was that current practice is consistent: pregnant women are almost universally excluded from enrollment in clinical trials. When women become pregnant during a trial, current practice is to immediately withdraw her from the study. There is an option for the PI to request continued inclusion on a case-by-case basis, as is true in general for study participants, but it was suggested that utilization of this option is rare. Some industry respondents indicated that although the woman is disenrolled, the study coordinators maintain contact with her and collect information on the outcome of the pregnancy. But others stated that not all pharmaceutical companies do this kind of follow-up as it is not required by regulation.

One of the participants had practical experience in conducting studies in pregnant women. She found two factors very helpful in that endeavor—safety data on the use of the product in pregnancy that may be available from government-sponsored studies and pregnancy registries, and decision-makers within the company who were familiar with the interpretation of drug safety data and the therapeutic needs of pregnant women.

Researchers and IRB members discussed the ethical principles involved when considering the enrollment of pregnant women. Nonmaleficence (do no harm) was cited as the main cautionary principle but justice (the inclusion of all groups who may benefit from the treatment) and autonomy (the right to decide for oneself) were mentioned. One IRB member stated that not all principles were weighted equally. He felt that nonmaleficence transcended autonomy—a perception apparently shared with others as reflected by the status quo. The FDA participant thought that initial studies that include pregnant women will likely be conducted in specifically identified therapeutic areas—perhaps making it easier to justify the risk by choosing well-documented and widely accepted areas of need. Both IRB and PhRMA participants agreed that the FDA guidance document on the subject will improve consideration of the topic and strengthen their policies.

1. Most companies exclude pregnant women from their studies.
2. IRBs generally have policies regarding the inclusion of pregnant women in clinical studies, based on the Code of Federal Regulations regarding vulnerable populations. They feel that their policies could be improved by the FDA guidance on this topic.
3. Some companies have experience doing clinical studies that include pregnant women. Information on drug safety gathered from other sources can be helpful in setting up clinical studies for pregnant women. Some women who become pregnant while enrolled in clinical studies may remain in some studies on an ad hoc, compassionate use basis.
4. FDA feels that studies should be done for certain products where the need is well established.

Question 3. Reasons for exclusion/barriers to inclusion

Rationales against the inclusion of pregnant women in clinical trials (found mostly in papers on reasons to include them) overlapped with the concerns raised by the industry stakeholders: nonmaleficence and litigation, enrollment issues, business concerns, and the lack of a regulatory mandate for their inclusion were all cited as rationales for exclusion in both the literature and among our stakeholders. Nonmaleficence and litigation were the most-cited rationales from both sources.

Question 3: Can you give me three or four reasons why a company or organization would not want to include pregnant women in clinical trials?

I can't think of three or four reasons why you'd <u>want</u> to include pregnant women.
—Pharma attorney

By far, the two predominant answers to this question were the desire to do no harm and the risk of litigation. Most participants mentioned one or both. Other commonly cited reasons to avoid including pregnant women were scientific validity issues, risks to drug approval and to company reputation, and the increased complexity to running such trials. Other reasons mentioned by one or two participants were the lack of advocacy for—or even awareness of—the need for their inclusion, the lack of a regulatory requirement or recommendation, and that it is not the historically acceptable way to do business (Table 5.1).

Do no harm: beneficence and nonmaleficence

the risk to the fetus . . . is a historically insurmountable hurdle.
—Pharma attorney

Stakeholders agreed that the most important deterrent to the inclusion of pregnant women in clinical trials is the fear of doing harm to a developing fetus. Most other considerations stemmed from that fear. Being overly cautious was a position that the industry has been very comfortable with.

I think it's first and foremost the ethical considerations of enrolling a woman when you presumptively don't have a clear sense of the potential teratogenic effect of product. So, historically, there's been a very strong reluctance to enroll pregnant women for fear of causing harm to the unborn fetus. I think that's the primary [reason]. I think that would be far and away the most important.
—Pharma attorney

Three prerequisites to drug testing in pregnant women were identified by the participants:

- Prior knowledge of the drug's safety for use during pregnancy
- Prior knowledge of its efficacy against the medical condition in question
- Prior knowledge of the proper dosing to achieve therapeutic benefit.

The problem lies in the fact that during a clinical trial, we don't know the safety of the drug, that's why we're doing the clinical trial. That's why we're doing the trial. So, under this kind of legal susceptibility, this volatile field, in the context of not knowing the benefit of the drug yet and not knowing the safety of the drug yet, then it makes sense that we would exclude this very susceptible population

TABLE 5.1 Reasons why pregnant women are not included in clinical trials.

Most frequently cited:

Do no harm

Litigation risk

Commonly cited:

Scientific validity issues

Risk to approval

Risk to company reputation/negative publicity

Business concerns

Occasionally cited:

Lack of advocacy for/awareness of the issue

Risk to drug market

Lack of regulatory requirement or recommendation

Not the norm/never been done

Lack of fetal consent

Areas of confusion:

Are there regulations against inclusion?

What is the extent of, and what are reasons for, medication use by pregnant women?

until at least the benefits are known. And we will not know the benefits of the drugs until the end of the phase III trials. We may think we know them but until we do the big clinical studies we don't really know the benefits of the drug, the true efficacy and benefits. So, exclusion is because we cannot strongly support that the benefit to the mother outweighs the risk to the fetus yet.

—Pharma physician

It's not just congenital anomalies or the effect on the pregnancy, the question is, what's the proper dose? Pregnant women get increased blood flow and hemodynamic changes that take place in pregnancy. I think we need to so some pharmacokinetic studies to make sure that the dose is the correct dose for pregnant women. If I expect efficacy, I want to make sure I have the correct blood levels to get that efficacy.   —Pharma physician

Do you even know how to dose it? I think [the anthrax experience] is a very cautionary example. You better check on your dosing if you're doing [trials in pregnant women].   —Pharma physician

Scientific validity: data interpretation

Pregnancy is just an outlier.   —Biotech physician

Drugs are tested in clinical trials to gain data from experience in enough people to result in statistically significant information to draw conclusions about the drug's efficacy, safety, and dosage for use in a general population. The people who are eligible to be included in the studies are fairly closely proscribed to exclude people who may make analyzing the data more difficult—people who have other medical conditions than the illness the drug is intended to treat, people who are taking other medications that may interact or interfere with the study drug, etc. A rationale for the exclusion of pregnant women articulated by the participants in the interviews was that pregnant women may complicate the interpretation of data if they were included in studies of drugs for the general population.

...they are so much of an outlier in terms of the normal physiology so they just exclude them. They exclude patients who have got too complicated a medical history, or who are taking too many concomitant meds. That's really basic.
                                                                —Biotech physician

It's the same thing that honestly drives really narrow patient populations in the studies they do anyway...   —Pharma physician

Pregnant women were cited as having a unique physiology that could impact drug studies in two ways:

- Drugs may affect pregnant women differently than nonpregnant women.
- Pregnant women may affect drugs differently than nonpregnant women.

Women's bodies are different during pregnancy and so you're not sure how that may skew the results of the drug or device being used on the woman. How [is] the [pregnant] woman's body going to affect how the drug is going to react?   —IRB member

It was felt that these factors need to be considered and that they should be evaluated in the context of a study designed for pregnant women rather than including pregnant women in a study designed for the general population.

Even if pregnant women were included in general population studies, the numbers enrolled would likely be low and would probably result in a lack of interpretable data to make recommendations for use of the drug in pregnancy.

You're not likely to get enough pregnant women to really draw conclusions. So ... if you allowed pregnant women to stay in a study, you will have really small number that really won't allow you to draw statistical conclusions because your sample size is so low. —Pharma physician

We want to make sure that research is always scientifically valid. If not, then you're putting people into research that does not have a possibility of having some benefit in the future, so then it's not ethical to include people in such a trial. —IRB stakeholder

Teratogenicity

Thalidomide has really scared a lot of companies. —Pharma physician

Enrolling pregnant women in research studies was seen by the participants as a complex and ill-defined process complicated by the fact that between 2% and 4% of pregnancies in the background population result in an infant with a major congenital anomaly. That percentage rises to about 15% when you include minor anomalies—those with no medical or only minor cosmetic significance. The causes of these anomalies are, in the great majority of cases, unknown, and those attributed to drug exposures are very low (about 2% of the 2%−4%).

Therefore, if you include pregnant women in studies, a certain number of infants will be born with birth defects just by chance or background occurrence. The risk cited by the participants is that the birth defects could be erroneously attributed to the drug exposure. The evaluation of the potential teratogenic effect of a drug on a fetus follows a well-defined scientific analysis that includes factors such as the gestational timing of the exposure in relation to the fetal development of the organ system affected, the effect in relation to the dosage, the consistency of the effect, etc. A teratologist will be able to provide a detailed analysis of the strength of the association between the drug exposure and the birth defect, but, especially when there are only a few cases, it is difficult to have complete certainty of causation. In fact, certainty of causation is easier to assure than certainty of noncausation. And despite the scientific evidence, causation could be attributed to the drug in the mind of the public, sometimes assisted by the legal community. This is what happened with the drug Bendectin (discussed in Chapter 1: Drug testing and pregnant women: background and significance), an effective and nonteratogenic drug that was removed from the American market because of litigation costs. Stakeholders disclosed that pharmaceutical companies are cautious about testing a new drug in pregnant women and risking its reputation in the marketplace if there is no mandate to do so.

Because, as you well know, bad things happen sometimes in pregnancies even when no drugs have been taken. So having one of those rare, but not clearly drug related events happening could cast a negative shadow on a drug forever and even prevent approval. So having one or two birth defects occur, even if they were background, you wouldn't have enough data to clearly say it was background, and it could really kill the drug.   —Pharma physician

I would not want the drug I was trying to develop to be tagged as being harmful to women who might become pregnant, [or] being blamed for spontaneous abortions or congenital anomalies. Part of the problem with enrolling just a couple hundred people is that you don't know how to interpret the data. You don't know if there's an association or not. The drug could be blamed for it one way or the other and I think that's unfortunate.   —Pharma physician

Regulatory rationales

It's just like a black hole.   —Biotech physician

There appeared to be some confusion about what local and international regulations allow and don't allow. Some respondents thought that FDA regulations barred the inclusion of pregnant women in clinical trials, others thought that European regulations did so. Some thought the human rights documents like the Belmont Report or the Declaration of Helsinki prohibited their inclusion. One respondent cited "different regulations from the time of the Helsinki Declaration until now not allowing ... the participation of pregnant women in clinical studies" as the reason for their being excluded.

Some participants stated that pregnant women were not included because FDA, or other regulatory bodies, do not require—or even recommend—that developmental drugs be tested for use in pregnancy.

Companies are not going to go there, quite frankly, because they're not regulated to go there and there are other special populations, like kids, that they're going to have to go to first.   —FDA member

Nobody even talks about it in your planning a study. It's just like a black hole. These days when you go [to FDA] for scientific advice when you're doing a program, you're looking at pediatric patients, they're pushing a lot for elderly patients, and it's not even on the radar screen about pregnant women.
                        —Pharma physician

Another regulatory issue is product labeling. Currently, because most drugs on the US market are not tested on pregnant women and animal study results may be ambiguous, they are labeled FDA Pregnancy

Category C, which states that, "human studies are lacking, animal studies have shown a risk or are lacking as well, but the potential benefit may outweigh the potential risk." Because pregnancy data are not collected for the purpose of adding information to the label, even when products have been on the market for years, most remain a Category C for the lifetime of the product. One respondent questioned the quality and usefulness of the data in the current labels and inferred that data collected from actually studying pregnant women would improve the information in the label and the ability to treat medically compromised pregnancies more efficaciously.

Business concerns

I'm glad this is anonymous.   —Pharma physician

Pharmaceutical companies are for-profit entities whose mission is to create and market drugs, biologics, and medical devices that prevent, treat, or suppress the adverse medical conditions that plague humanity. That mission includes making a profit in order to compensate the people who work to achieve the mission and to spur the continued research required to sustain innovation and grow profitability.

For the pharmaceutical company, the ultimate purpose of the clinical trial is to confirm the efficacy and the safety of the new compound in order to get their product approved as quickly as possible. Years of research and millions of dollars have already been spent in shepherding the potential product to the clinical trial stage.

Companies want to get their studies approved as fast as possible, they don't want any extraneous issues that could go wrong.   —Biotech physician

Industry constantly thinks about the risk to the drug.   —Pharma physician

The conduct of clinical trials in populations that are peripheral to the primary target population is a secondary concern. On this topic, participants stated that,

They're not going to go there until they know this drug is going to make money for them.   —FDA member

They don't want to put their drug in a position where the drug may receive an unfavorable review from the FDA or any regulatory party.   —Pharma physician

In the context of business, several participants named adverse notoriety for the company as a reason to avoid testing the drug in pregnant women. "[I]f something were to go wrong and people found out, 'wow, you were testing this drug on pregnant women...,'" the impact to the company's reputation could be significant and difficult to recover from.

> *It's not just the fear that something can go wrong − I'm glad this is anonymous − [but] when you consider that things can go wrong in a clinical trial and the most published news about clinical trials is the negative information ... [y]ou conducted a clinical trial and something goes horribly wrong, the name of the facility is put out there, the name of the physician that conducted it is put out there, the IRB that reviewed and approved it is put out there ... even as careful as you can be...*   —Pharma physician

The risks of research and pregnancy

> *It's risky research.*   —IRB member

Some stakeholders gave me the impression that they perceived the risk to the pregnant woman and the fetus from participation in *any* clinical trial to be extremely high. They did not consider it on a case-by-case or a trial-by-trial basis, but thought the risk to be very high across the board.

> *Extreme safety risks − for the mother and the unborn child.*   —PhRMA attorney

And they suggested that pregnancy itself was a high risk condition:

> *Not only is the woman's body different, [but it is] potentially more vulnerable health-wise while pregnant...*   —IRB member

But does everyone agree with these statements? Research has shown that pregnant women and health care providers overestimate the teratogenic risk of drugs and environmental factors. Do researchers also overestimate the risk of pregnant women participating in clinical trials? Also, are women less healthy and more vulnerable when pregnant? Is pregnancy a disease state or a healthy state? Are pregnant women, as the Common Rule suggests, a vulnerable population?

The informed consent of the fetus

> *The child has no voice.*   —IRB member

A couple of the participants suggested that one of the reasons why pregnant women are not included in clinical trials is because it is impossible to consent the fetus.

> *There's not just one person, there's two people at risk. You have your second person at risk that has no voice whatsoever. You have the mother who can say, yeah, I think I want to do this, but when she says that, she's speaking for a child as well, and the child has no voice. I think that's the hardest part.*   —IRB member

Whatever you think of the moral status of the unborn human life, medicine can treat the fetus ... as a patient and the law tends to as well. So in practice there's a human being there, who in terms of human subjects protection is by definition vulnerable. —IRB member

In responses to the questions, it was suggested that the fetus is an entity whose needs, in some regards, should be considered as independent from the pregnant woman. This fetal status could be considered as a barrier to the inclusion of pregnant women in clinical trials.

Limit testing to drugs indicated for use by pregnant women

[T]here's actually a risk of non-treatment to the fetus as well.
—Pharma attorney

I can't think of 3 or 4 reasons why you'd want to include pregnant women — unless it's a situation where you have a specific case where you need to study your intervention in the setting of pregnancy because pregnant women are going to get your drug in the (post-marketing environment). —Pharma attorney

This response suggests that pregnant women will only use a drug that is intended for use by pregnant women. We know that many fetal exposures are unintended—the use of medication by women who do not yet know that they are pregnant. We also know that many medical conditions are not specific to pregnancy but may occur in pregnant women (heart disease, stroke, diabetes, infections, asthma, lupus, Crohn's disease, etc.).

Therefore it is difficult to determine with any specificity which drugs in development will or will not be used by pregnant women, unless they target conditions exclusive to men. The prudent assumption would be that most drugs, if they are effective, will be used by pregnant women.

Since it would be impossible to know what drugs may be used by pregnant women in the marketplace, are there others ways to target research for them?

I doubt that there's going to be much interest in sponsoring clinical trials for the use of chronic meds for non-life-threatening conditions or where there is a reasonably well-established treatment paradigm. I mean, you have insulin for diabetes, you suffer with your symptoms for allergic rhinitis et cetera, et cetera. A lot of these, you can kind of manage through, but there's others that as a pregnant woman you can't always wait. And there's actually a risk of non-treatment to the fetus as well. Then I think you have a much more compelling ethical argument for experimentation. —Pharma physician

Lack of advocacy

There's not been anyone to advocate for it.   —Biotech physician

There has been no push for a change, no pressure on industry to do this. The lack of experience, the perceived hassle, the increased complexity of the study design, IRB resistance, legal considerations, all conspire to maintain the status quo.

It's not very high up in the consciousness of most people conducting clinical studies.   —Biotech physician

I think there's a long history of not doing it, so trying to get over the inertia of doing that is very difficult.   —Biotech physician

And yet, there are suggestions, like the draft FDA guidance and the Second Wave Consortium, that advocacy has begun.

They have to get over the sort of natural reaction of, 'oh, boy, we really can't do this' and then get down to the fact that, 'yes, we can do it, how are we going to do it?'   —FDA member

Key findings for Question 3

One of the aims of this project was to isolate the concerns that are articulated by the pharmaceutical industry and their IRB and FDA colleagues regarding the inclusion of pregnant women in clinical trials. In order to fully evaluate the issue and to potentially devise means to alter the status quo, a better understanding of the potential barriers perceived by this powerful stakeholder is needed.

1. The fear of causing harm to a fetus was cited as the most important concern limiting the inclusion of pregnant women in clinical trials.
2. The fear of litigation is one of the major concerns that is limiting the inclusion of pregnant women in clinical trials.
3. The efficacy, safety, and proper dose of a medication must be known to some extent prior to testing the drug in pregnant women.
4. Industry has little experience designing clinical trials that include pregnant women. More information is needed to assist with the design of such studies.
5. National and international regulations regarding the inclusion of pregnant women in clinical studies are not well understood.
6. Studying drugs in pregnant women would provide valuable information for the label, which would improve the treatment of pregnant women.

7. Industry is reluctant to risk the approval of a drug for the nonpregnant population or the reputation of its company by testing drugs on pregnant women.
8. A sufficient number of pregnant women must be included in a study of pregnant women to ensure that the data collected is interpretable.
9. Industry perceives little motivation or advocacy for the study of its products in pregnant women.

Note

1. FDA's Guidance Document To-Do List. *The pink sheet.* (2011, January 3). Retrieved from <https://pink.pharmaintelligence.informa.com/>.

Chapter 6

Perspectives from the industry: on inclusion

Who's really advocated for clinical trials in pregnant women?
—Biotech physician

Advocacy

Question 4. Advocacy awareness

Are you aware of the US Food and Drug Administration (FDA) guidance? Have you heard of the Second Wave advocacy group?

In 2012 most stakeholders were not aware that a draft FDA was being prepared at the agency. Obviously, the FDA staffer was aware, but the Pharmaceutical Research and Manufacturers' Association (PhRMA) stakeholder was not. A couple of participants [one in an Institutional Review Board (IRB) and one in biotech] had learned of the draft guidance because its title was included on the FDA guidance to-do list published in "The Pink Sheet" in January 2011.[1] Two industry employees (one biotech and one pharma) questioned whether the draft FDA was working on was a new iteration of a previous draft they had heard about "some years ago."

The FDA stakeholder confirmed that the draft guidance is "working its way through the system of review and clearance," which, she stated, can "move ponderously slowly."

> *... there is a growing sense of comfort in the idea that one can do these studies, given the number of studies that ... have been done in antivirals and HIV, and to some extent, the onset of concern around medical countermeasures. You see the light bulbs go on when you talk about the ampicillin/anthrax issue. That's been a hallmark set of studies to really wake people up ... Pregnant women may not be just like women of the same age who are not pregnant when it comes to medicines, and how are we going to deal with it?* —FDA member

The ampicillin/anthrax issue, previously discussed in Chapter 1, Drug testing and pregnant women: background and significance, refers to the 2002

Pregnancy and the Pharmaceutical Industry. DOI: https://doi.org/10.1016/B978-0-12-818550-6.00006-4

American College of Obstetricians and Gynecologists recommendation of the use of amoxicillin to treat pregnant women who were potentially exposed to anthrax. The dosage and frequency regimen was found in subsequent studies conducted several years later to be ineffective in the treatment of pregnant women. The implications of this finding highlight the dangers of generalizing data from nonpregnant subjects to pregnant women and the need for research on the actual target population.

Most stakeholders had never heard of the Second Wave Consortium. Of the two participants who had, one was at FDA and one worked at a biotech company. Both said they had heard mention of the group by a speaker at a conference. The FDA interviewee thought that the Second Wave Consortium was doing important work in getting the issue into the public domain.

> *Because of the sea change that we have to have here, and because there are so many stakeholders in this conversation, this is how we're going to get started.* —FDA member

Other participants concurred that there is a lack of awareness of the issue that may hamper efforts to change. One said,

> *[T]here's just been no interest in looking at this. There's not been anyone to advocate for it, right? So, it's like the difficulty with pediatrics. Over the longest time there was just not really sufficient advocacy for the position from the right stakeholders, until there was some financial incentives. This is even starker.... Until you told me about the Second Wave, I hadn't heard about anything.* —Biotech physician

Key findings for Question 4

Awareness of the issue was seen as a prerequisite to advocacy, the lack of which was consistently cited as a barrier to change. Most of the participants were not aware that the issue is being debated in some quarters, nor were they aware that FDA has a guidance document in draft. Lack of awareness on the part of the general public was also identified as a critical issue, particularly because, as one participant stated, a "sea change" in thinking will be required to address it. This dearth of awareness at all levels was seen to be a potential barrier to the initiation of change and a facilitator of the status quo.

1. Three opportunities were identified that could facilitate both awareness of the issue and its potential resolution: (1) public and government interest in bioterrorism (the "anthrax issue" was cited as an ideal illustration of the problem), (2) prior success in conducting clinical trials in HIV positive pregnant women, and (3) the success of the effort to include pediatric patients in the drug development process.

2. There is a lack of awareness among industry employees and within related organizations about the issue of the inclusion of pregnant women in clinical trials, about the impending (at the time) release of an FDA guidance document on the topic, and about the Second Wave Coalition advocacy group. The stakeholders implied there is also a lack of awareness of the issue among the general public. The implication of this dearth of awareness at all levels was seen to be a potential barrier to the initiation of change and a facilitator of the status quo.
3. It is critical to get the issue into the public domain in order to change current thinking and get stakeholders involved. The work of the Second Wave Consortium was felt to be important in this regard.
4. There is opportunity for change utilizing the government's and the public's interest in protection against bioterrorism.
5. There is opportunity for change utilizing the work that has been done with pregnant women in clinical trials for HIV treatment and for the prevention of vertical transmission of HIV to their fetuses.
6. There exist similarities between the exclusion of pregnant women from clinical research and the former exclusion of pediatric patients from clinical research. There may be lessons learned from the endeavors of the pediatric sector that have resulted in mandated pediatric clinical studies.

Stakeholders on inclusion

Question 5. Inclusion

Aside from studies that are specifically about conditions of pregnancy, can you give me three or four reasons why a company (or an IRB) should or might want to *include* pregnant women in clinical trials?

> *You mean aside from the fact that it's the right thing to do?*   —FDA member

It is not for lack of understanding of the need for data that pregnant women have not been welcomed into clinical trials. All of the stakeholders I interviewed could speak to the need for a better understanding of how pharmacotherapy can be employed during pregnancy in general and how to obtain the data needed to inform prescribing information for specific products (Table 6.1).

Medical need

The reason most often cited for the need to include pregnant women in clinical trials was to provide treatment for pregnant women who have medically compromised pregnancies. All of the stakeholders could answer the question of "why," it was the question of "how" that was difficult.

> *You're flying blind when pregnant women get sick.*   —Biotech physician

TABLE 6.1 Reasons for including pregnant women in clinical trials.

- It's the right thing to do
- There is medical need
- When the benefit to the woman/baby exceeds the risk to the woman/baby
- To help health care providers treat their pregnant patients correctly
- To provide accurate information to the product label
- To aid the company reputation
- To fully evaluate the product's safety profile
- To develop medicines that treat the pregnant population
- To improve insurance coverage for medications
- As a competitive advantage for the company
- To emulate best practices in other special populations like the elderly and pediatrics

I think we have an unmet medical need which is, we don't know enough about these drugs to feel comfortable using them in pregnant women and therefore women who really need therapy for whatever conditions they have that is concurrent with their pregnancy, you have no information. So you might have women either taking a drug that they don't know enough about, or not taking a drug that they do need, and either way, we have no information to guide care. So guidance of care is a very important reason.   —Pharma physician

There are about 4 million births in this country each year, so there are a lot of pregnant women out there, and a lot of women in their childbearing age that might get pregnant. And many of these women have chronic diseases that need treatment throughout pregnancy for serious diseases during pregnancy.
            —Pharma physician

How are we going to move forward? How are we going to be able to help those who need it if we don't conduct trials? The whole reason for research goes back to—it's the same question—we need the information.   —IRB member

If we were studying a drug for a condition that had a high rate in women who might be pregnant, for example, diabetes, or a large number of people at risk for a condition where you might actually be treated before you became pregnant, you need to know what would happen. Should you stop the drug or can you continue it? The same kind of thing with HIV. You might be treated before you even get pregnant. You have to know can you continue it? So for anything that has a reasonable prevalence rate in women who are likely to get pregnant, it probably is worth studying—but maybe not before you get the original approval for the product—that's a different question. I don't think it's a box check, where every drug should be tested. It should be for a drug where the condition might reasonably occur in pregnant women. —Pharma physician

So if we consider that the segment of the pregnant population is an important population, and we know that women who get pregnant sometimes need to use medications, either because they have a current condition or because they develop a condition during pregnancy, and they need the medication, we need to know of the safety of the medications. We need to know. Because those women, when the drug goes to the market, those women will need a medication and use the medication, and if we don't have information then we are putting that population at an even higher risk. —Biotech physician.

If the benefit exceeds the risk

Sometimes we do know from doing animal models that the drug is safe ... We know from phase 1 and phase 2 studies that the drug does have benefits, it's an effective drug. We have a pretty clear idea that it's safe and effective. And in the context of the higher benefits to the mother than the risks to the fetus, then in that context, there's no real reason to exclude these women. —Biotech physician

Company reputation

I think also the customers who use the product, if they know that the drug company is really diligent about monitoring the safety profile of the product in pregnancy, then I would personally be more likely to take that company's drug over a company that didn't do it. Because I would feel that the company that conscientiously really tries to evaluate it are the kind of people I'd like to take their drug and not companies who are not interested in evaluating the drug. The ethical things are ... good PR for the companies and good [for the] reputation of the company. —Pharma physician

To provide information for prescribing health care providers

If your drug has a possibility of being used in pregnancy, you would want to have at least some body of data in the company database so that ... you could advise physicians and clinicians when they are asking questions about whether it's safe to give that ... drug to a pregnant woman that is a patient of theirs. —Pharma physician

I really think we've got to the stage whereby evidence is so much better than— I always go back to the evidence—it's so much better than guessing. —Pharma physician

Good data ... will reduce the risk for providers. We have to remember that they assume a certain amount of risk with dispensing or prescribing a drug that might not have a lot of information. —Pharma physician

To fully describe the safety profile of the product

The one area that I think that companies should include pregnant women in clinical trials is in vaccines because I think it's important. Like with the flu vaccine, a lot of pregnant women get the flu vaccine to protect their baby and also to protect themselves but nobody ever does studies in pregnant women. I don't think they do studies in pregnant women with vaccines, I haven't seen any. Pregnant women were at huge risk, they were the ones who were dying with the swine flu epidemic. So I basically think that if you don't include pregnant women in your patient population and they are a population who is likely to receive the product, the vaccine, the drug, whatever, you have not adequately described the safety profile of your product. —Pharma physician

To develop medications to treat the population

I can come up with multiple reasons but the big one is so you can develop drugs that are actually known to be efficacious to use in pregnant women or understand how they may affect the fetus so I think it all just sort of relates to trying to derive a benefit for these pregnant women. I think that reason is, in and of itself, why we conduct research in general. We do research to further science that will hopefully benefit various populations to treat diseases and conditions and that's the whole reason why there is research and why we are developing new medicines. —IRB member

To inform labeling

[The information] "*can help you to work with the FDA or some regulatory body as to where to start in terms of how to categorize the drug in terms of FDA pregnancy category.*" —IRB member

And if we have no information as to such simple things as pharmacokinetics in that population, are we providing the best prescribing information for use of the drug in pregnancy? Whether it's formally approved for use in pregnancy or not, but just knowing that it could be used in pregnancy. —Biotech physician

To improve insurance coverage

One other reason may be to inform insurance companies because sometimes insurance companies have an influence on whether they will pay for a drug or not. Maybe this information [will] inform them so that they didn't disqualify the

drug or tell a woman that she had to pay out-of-pocket because they wouldn't want to be involved in any potential liability or risk.   —Pharma physician

For competitive advantage

A couple of participants rather reluctantly raised the idea of competitive advantage but did not place it high on their lists of reasons to do the studies.

Well, I don't know if I really want to raise it, but it could be a competitive advantage ... over another product, if you ... see that you're effective in an area and have a better safety profile over a competitive product. I don't like to use safety for competitive advantage so that I'm kind of reluctant to make much of a case of that. FDA is very cautious about using safety as competitive advantage [too].   —Biotech physician

To treat pregnant women like other special populations

I don't think we should see pregnant women as something different. Pregnant women should just be seen as part of the patient population, as elderly are seen that way.   —Pharma physician

I think it's maybe a similar approach to the way people thought about pediatric trials a while ago, which is that we don't want to risk harm to a child so were not going to use study drugs in them. And then it became apparent that, neverthe-less, children needed treatment for conditions and so many physicians were pre-scribing products without an adequate understanding of the risks and benefits. And it became apparent that, from a society point of view, [we need] to actually conduct the trials and figure out what products worked and what didn't. So I think it's the same thing there, if you get past the ethical hurdle, you do have this issue of knowing that pregnant women, for some conditions, need to be treated. And that if you know that women need treatment of some sort, there's an argu-ment that you may as well figure out what treatments are safe and appropriate. I think that's only one reason—but I don't know what the other three or four would be. Maybe I'm too stuck on the ethics piece.   —Pharma attorney

Key findings to Question 5

1. Members of pharmaceutical companies, IRBs, PhRMA, and FDA, physi-cians, lawyers, and business people, agree that there are compelling rea-sons to conduct clinical trials in pregnant patients based on the need for information on how to treat them effectively.
2. Conducting trials on drug treatments for pregnant women is advantageous for the pregnant women, the health care providers, the prescribers, the FDA, the pharmaceutical company, and society in general.

3. Pregnant women and their fetuses are at a higher risk of adverse medical consequences if they are not included in clinical trials than if they are included in clinical trials.
4. Increasing the inclusion of pregnant women in clinical studies is such a difficult and controversial undertaking that any and all suggestions for how to make it happen should be on the table for consideration.

Question 6. Safeguarding safety

If pregnant women were enrolled, what steps could be taken to safeguard the fetuses and the pregnant women who consent to participate?

[I]t is difficult for industry [to make these] very difficult ethical considerations when it's a subject of so much external criticism. —Pharma attorney

In many but not all cases, participants answered this question in previous discussions and so were not asked it again during the interview process. However, some additional thoughts were recorded on this subject.

Steps that were recommended to be taken *before* initiating a clinical trial included:

- an evaluation of the alternative treatments available to make sure they would not be more appropriate,
- a thorough review of the animal toxicology data,
- an understanding of the pharmacokinetics (PKs) and pharmacodynamics of the drug,
- knowing if it crosses the placenta,
- knowing what percent is protein bound, and
- looking at other products in the same drug class or with a similar molecular structure.

During the course of the study, researchers should:

- perform additional fetal monitoring,
- schedule frequent study visits,
- involve specialists in maternal-fetal medicine in the design and conduct of the study.

Recommendations were made to keep evaluating the data on an ongoing basis being alert to any signals that might be meaningful.

You have to keep looking at this—whether it's results of prior clinical studies, results of observational data, as you keep looking at this and refining over and over again, if you're planning a new study or if you're enrolling the next patient into the next study. Each pregnant patient entering the study should be able to benefit from the knowledge gained from every patient that has gone before her in this type of the setting. —Pharma physician

An industry lawyer recommended an independent oversight group he described as "sort of a super IRB that can be maybe an extra powerful data safety monitoring board . . . that would be the one to make these difficult ethical decisions." He said he did not think industry would engage in this kind of research without one.

Key findings for Question 6

Suggestions about how to minimize risk were offered for consideration. Many participants felt that there are subject matter experts within and outside of industry who could be consulted about how to design a trial for pregnant women that would safeguard the woman, the pregnancy, and the fetus.

1. Data collection and analysis should be applied in an iterative fashion so that each pregnant patient entering a study should be benefit from the knowledge gained from every patient that has gone before her.
2. A pregnancy-specific independent data safety monitoring board should provide oversight and decision-making functions.

Alternative pathways

Question 7. Other opportunities

If not enrolling in clinical trials, what are alternative ways to get this information? Are there alternative study designs or data collection methods that could include pregnant women?

> *Could we do better at collecting the data from pregnant women who are taking drugs that are on the market? I'm sure we could.*   —Pharma physician

Preclinical

Interview participants believed that more could be done during animal testing to obtain data on the PKs and pharmacodynamics of compounds in the pregnant state. Also, it was suggested that new modeling techniques should focus on the difficulties of getting information on the effects of exposure in pregnancy. One pharma attorney said, "You have, increasingly, modeling of various types. I guess you have animal models to some extent. None of those are really adequate substitutes but that work goes on now."

Clinical

Several of the interviewees thought that inadvertent pregnancy exposures during clinical development are a potential source of data that we have traditionally neglected. One participant asked, "Do we need to change our approach to inadvertent pregnancies in clinical trials in order to capture data?" Many shared her opinion that "we maybe don't do as good a job as we should in taking advantage of the circumstances."

Others agreed and added that it's not only an opportunity to collect safety data, but it's also "an opportunity to possibly gain some PK data. I think it's a clear opportunity and the more you get prepared for it the more likely you can take advantage of it." This participant was recommending that some inadvertent pregnancies should be expected in any study and the protocol should include a procedure to take advantage of the learning opportunity. Are there preexposure blood samples that can be compared to other samples taken at different gestational weeks of exposure during the pregnancy—whether or not the woman is maintained in the trial? This kind of valuable information would be very difficult to obtain outside of a clinical study.

> *If a woman becomes pregnant in a clinical trial, we can ask her if she would volunteer to be in a pk study, as a voluntary thing, not a mandatory thing. 'Would you mind giving us a sample of your blood? Just to know how much of the medication we should give you.' That would be such an amazing source of information.*   —Pharma physician

Currently there is no mandate saying that a woman who inadvertently becomes pregnant during a clinical trial must be continued or discontinued from the trial. Most, however, are immediately disenrolled. "I think the withdrawal instruction is pretty strong," said a pharma member, but added that "you can readdress restarting the drug" as an exemption to the protocol on an ad hoc basis following appropriate analysis and agreement from the principal investigator (PI), the Data Safety Monitoring Board, the IRB, and the Sponsor. But, whether or not the patient's condition warrants remaining in the trial, many interviewees agreed that the pregnancy—which has likely been exposed for several weeks—should be followed to outcome.

> *Following a pregnant women from clinical trials is not mandatory. So that's the first thing that I think all companies should do—that the woman was pregnant in the clinical trials, even if she gets excluded from the trial, from now on they should be followed up to know the outcome. That's rule number one.*
>   —Pharma physician

> *Right now, we're missing, we lose that data, we're not following it. I'm sure you've been in conversations over the years at the company where the argument is ... 'well, yes, we might have them, but we're only going to have four, five or six. We can't do anything with that data, so were not going to collect it.' And I think we have to change that mindset, appreciating fully that you can't use traditional kinds of analysis on four, five or six, but that information is important.*   —Pharma physician

However, another physician cautioned, "What is probably difficult to expect industry to swallow, would be some sort of commitment to follow-up a child for the rest of its life."

Postmarketing surveillance

Once the product is approved and goes to market, there are further opportunities to identify and capture pregnancy-exposure data.

> *It becomes a pharmacovigilance issue. I think this whole thing probably fits better in pharmacovigilance that it does in drug development.*　—IRB member

> *Perhaps doctors who are treating their patients might be willing to voluntarily submit data—with the patient's permission because of HIPAA—but maybe through some sort of, it's not really a trial but some sort of process could be developed ... At least they could have sort of a databank of data that they could refer to that could be analyzed, perhaps data that could be open to them to also receive.*　—Pharma physician

It is important to note that HIPAA (the Health Insurance Portability and Accountability Act) includes an exception for health care providers reporting safety surveillance data. "Covered entities may disclose protected health information to a person subject to FDA jurisdiction ... for public health purposes related to the quality, safety or effectiveness of an FDA-regulated product ... includ[ing] ... collecting or reporting adverse events ... conducting post-marketing surveillance."[2]

Voluntary reporting to regulatory agencies of adverse events experienced by women taking approved medication is suspected to be very low. One paper from New Zealand found that fewer than 9% of maternal and perinatal serious adverse events were reported.[3] Decision-making relying solely upon spontaneous reports of adverse experiences that occur in the postmarketing environment is ill-advised.

Regulatory agencies in major markets around the world (e.g., the United States, EU countries, Japan, etc.) require that sponsors requesting approval of new medications identify the risks and side effects found in its preclinical and its clinical studies. In addition, a description of the steps that the sponsor will implement to identify, evaluate, and report newly recognized safety issues once the product is on the market is also required as part of the application process. Since 2002, FDA mandates that sponsors include a section addressing 'special populations' that includes pediatrics, geriatrics, and pregnant women. In it, the company must address how the company plans to monitor the use and safety of the new drug in these populations. To monitor use in pregnancy, such steps could include special clinical studies, epidemiologic studies, and routine or enhanced surveillance.

I think the Swedish pregnancy registry and the Merck pregnancy registries are great ideas, great programs. I think that in risk management plans they should always be included. —Pharma physician

Pregnancy registries

Pregnancy registries were the most frequently cited method of collecting pregnancy exposure information.

The registry is the primary way to systematically collect that information. —Biotech physician

I feel as though it should almost be like, 'why shouldn't you do a pregnancy registry?' instead of like, 'you're going to have to do a pregnancy registry.' It should be the minimum. —Pharma physician

There are different types of pregnancy registries—population-based registries like the Swedish Medical Birth Register; company-based pregnancy registries like those run by Merck and other pharma companies, including collaborative registries like the Antiretroviral Pregnancy Registry in which multiple companies are involved; and independent registries run by academic medical centers, like the Anti-Epileptic Drug Registry. FDA maintains a list of pregnancy registries on its website.[4]

The study interviewees had various experiences with pregnancy registries.

We've done a few nice little observational studies using the Swedish registries where for the cost of some statistical help, the registries are all available. —Biotech physician

… there was a pretty good study from one of the teratology consulting groups from Israel but part of a European group, where for every women who called in with a question about an ACE-inhibiting drug they'd say, 'okay, we have a study, we would like to follow you long-term,' so they began filling out the forms and created it. That's a relatively inexpensive way to collect some long-term or at least pregnancy outcome data with people who self-identify. —Biotech physician

However, they also pointed out some of the methods' limitations.

The trouble is, [registries] … take more time than, say, if you did a clinical trial. [T]he difficulties with registries … are, of course, identifying the patients, making sure they get enrolled in the registry, getting the results into the registry and so forth. I guess the question would be, do you think that they get the data in a timely enough manner or that a trial would get … the data probably a little earlier? —PhRMA member

It's unfortunate ... that it's a teeny tiny fraction of the women who are exposed to a particular compound, actually register. Even then only a subset of them are actually followed to the conclusion of the pregnancy so I think better use of the registry approach would be a great thing. —Pharma attorney

Another idea generated by a study participant was the use of the pregnancy registry methodology—studying pregnant women who self-identify or who are reported by their health care provider in the postmarketing environment—for a PK study.

We could have the Phase IV trial ... a single arm kind of trial. I would prefer to call it a study rather than a trial. For example, you could establish a pregnancy registry or surveillance program ... and when women call up and say I just became pregnant, we can ask, 'would you like to be in a study where we will take one [blood] sample? In exchange, we will do a checkup for your baby, or we will take a picture of the baby for you and it will be framed, or a magnet for the fridge,' and then if they accept that—the context is you are already a registered participant who wants to volunteer, that's what I would do for those PK studies. We know that we don't need to have hundreds of women for these studies, a few women and you can do the entire study. I would do it in the context of a pregnancy registry or surveillance program, only those voluntary studies. —Pharma physician

So the pregnancy registry approach appeared to be an accepted means of obtaining safety data—albeit with some room for improvement.

Epidemiological studies

In addition to a pregnancy registry, participants suggested the use of other studies in the postmarketing environment utilizing electronic medical records or other databases containing information on drug exposures and pregnancy and fetal outcomes. They suggested that this is an emerging field where possibilities for improved data capture and analysis are being developed.

I would use the spontaneous reports to see if there's an issue and only if the spontaneous reports are suggesting something, would I then try to design either a registry or an observational study—because they're a lot of work unless you know what you're looking for. —Pharma physician

Do a large database study using electronic medical records to identify what drugs are taken and what congenital anomalies are identified, but they're not easy to do and they're not cheap to do. —Pharma physician

Company and regulatory agency support

One of the interviewees, a physician who had worked at several pharma and bio companies, expressed frustration with the lack of attention given to the issue within the companies. She suggested both an increase in sensitivity and knowledge within the companies and requested assistance from the regulatory agency on data review and analysis.

> *I was seeing signals and I would take it to our safety review team meetings at [the company], and everyone was, "no." They really were quite dismissive because these are all small studies. So I think being able to have people who are in the companies who are educated with regard to pregnancy and pregnancy data and how to get pregnancy data. And trying to standardize the review, getting not just the companies reviewing the data. I'm getting to feel as though it's very ad hoc at the moment, the way we approach pregnancy. Maybe there could be a standardized way of looking at pregnancy data, especially from a post-marketing point of view. Some guidelines like that. When is there a safety signal that you should be doing more? Does it have to be statistically significant or what? That kind of stuff.* —Pharma physician

Key findings for Question 7

Stakeholders were able to articulate alternative methods of obtaining pregnancy-exposure data—many of which are currently being used to some extent, like spontaneous reports and pregnancy registries. Most focused on methods in the postmarketing environment—after the drug is approved and on the market—because of the higher comfort level with pregnant women using drugs whose safety and efficacy is more well-established. Some suggested making better use of opportunities to gather information from women who become pregnant during clinical trials. One pharma physician conceded that, "Clinical studies alone may not be adequate, particularly for chronic exposures."

1. There are opportunities to improve our knowledge of the efficacy and safety of medication use in pregnancy in preclinical techniques and analysis, in inadvertent pregnancy exposures during clinical trials, and in postmarketing surveillance, pregnancy registries, and epidemiologic studies. Current methodologies could be improved and new methodologies should be explored.
2. Pharmaceutical company support and funding for the collection and analysis of use-in-pregnancy data would be helped by an articulated medical and societal perception of need, by company commitment, and by regulatory agency pressure.

Notes

1. FDA's Guidance Document To-Do List. The pink sheet. (2011, January 3). Retrieved from <https://pink.pharmaintelligence.informa.com/>.
2. U.S. Department of Health and Human Services. (2017). *Health information privacy: Public health.* Retrieved from <https://www.hhs.gov/hipaa/for-professionals/special-topics/public-health/index.html>.
3. Farquhar, C., Armstrong, S., Kim, B., Masson, V., & Sadler, L. (2015). Under-reporting of maternal and perinatal adverse events in New Zealand. *BMJ Open*, *5*, e007970. http://doi.org/10.1136/bmjopen-2015-007970.
4. U.S. Food and Drug Administration. (2018). *Science & research: Pregnancy registries.* Retrieved from <https://www.fda.gov/ScienceResearch/SpecialTopics/WomensHealthResearch/ucm251314.htm>.

Chapter 7

Perspectives from the industry: on litigation, regulation, incentives, and indemnity

I think it would take a woman CEO.   —Pharma physician

Litigation

The 2009 Second Wave Consortium workshop identified a missing piece of information for which this project sought clarification: "How influential is the industry's perceived risk of litigation—and is it a real risk?"

Question 8: Litigation risk

Question 8: Do we know that allowing pregnant women in clinical trials would result in litigation or are we presuming it would?

> *The elephant in the room is litigation.*   —Biotech physician

In answer to this question, most respondents said that they presumed that liability would increase if clinical trials were conducted with pregnant women, but they were not sure that it would. Since there is little experience with clinical research in this population, we really do not know.

> *...when you talk to the OB people, one of the reasons why the liability is so high is that we get blamed for all the abnormalities that could occur,... I hear that there are a lot of malpractice suits concerning congenital anomalies...*
> —Pharma physician

Of the two participants who said they know that litigation would increase, one said he knew it because of reports in the media and the other knew it from personal experience at her company. The latter's experience involved an approved product that was later found to raise the risk for certain birth defects.

Pregnancy and the Pharmaceutical Industry. DOI: https://doi.org/10.1016/B978-0-12-818550-6.00007-6
105

One pharma company employee said, "people are suing already when we are excluding them." When asked to explain, he said that his company has pending litigation concerning the exclusion of a woman from a trial during which she became pregnant and one concerning a pregnant woman who was prevented from enrolling in a clinical trial, because the company was "not providing them with the drug that they think is necessary."

A representative from Pharmaceutical Research and Manufacturers of America (PhRMA) stated that, "there are not a lot of lawsuits filed with respect to clinical trials" in general. This is corroborated by the medical literature which indicates that, "the risk of incurring liability during the early stages of drug investigation is actually quite small whereas the potential for substantial liability is much greater once a fetotoxic drug enters widespread use."[1]

The issue of informed consent was raised by several participants such as a pharma physician who said, "if they had informed consent, I can't really see a huge risk of litigation versus other studies that we do."

You're following the regulations, you obtained IRB approval so it's been considered from an ethics perspective, the person's been informed about it, the risks have been minimized as much as possible, and you're doing it to help pregnant women right there in the trial or in the future. —IRB attorney

A potential increase in litigation, the lawyer continued, "should not be a reason to stop people from including pregnant women in clinical trials. I don't think there's going to be that much of a boom in litigation for the industry."

Citing the anthrax study, an Institutional Review Board (IRB) lawyer stated that, "pregnant women are being included in this trial for a very important reason just like people who are not pregnant are included in clinical trials."

I can count on one hand the amount of calls that come to me where someone was actually damaged or injured in a clinical trial and they needed an attorney. That's over 15 years. —IRB member

Not everyone agreed.

The decision to sue is something that the company can't control. I would make sure the informed consent is as strong as it can be and Investigator's Brochure contains disclosures of all data to date about risks. It would be a benefit/risk analysis. We can defend on causation, bring in experts, particularly to discuss the science behind the defect. But when playing to a jury—I have [children]— any juror might see the case as a parent with a child [would]. So I think the litigation risks are higher. —Pharma attorney

Key findings for Question 8

Because birth defects occur at a rate of 3%−4% in the general population, birth defects would therefore be likely to occur in 3%−4% of the infants born to women who participated in clinical trials. The expectation among many participants was that litigation would follow these adverse events. However, it was acknowledged that, since we have little experience with pregnant women in clinical trials, we really don't know.

1. There is a perception that the risk of lawsuits against a company would be higher if drugs were being tested on pregnant women. But, because we have little experience in this area, we don't know if the litigation risks would be higher in clinical trials with pregnant subjects than in clinical trials in general.
2. There is a perceived risk that excluding pregnant women from clinical research could result in litigation due to adverse pregnancy outcomes caused by (1) restricting pregnant women from getting the drug they needed, or (2) caused by a drug that was not fully evaluated when it was put on the market.
3. Thorough informed consent, complete disclosure in the Investigator's Brochure, US Food and Drug Administration (FDA) approval, IRB review, risk minimization activities, and the disclosure that the trial is intended to help pregnant women now and in the future could help protect the company from lawsuits in clinical trials that include pregnant women.
4. Our litigious society, the emotional component in jury trials, and increased litigation risk in the obstetrical community in general could result in an increased risk of litigation in clinical trials that include pregnant women.
5. The risk of liability for injuries that occur during research in general is low.
6. Some respondents, including company lawyers, believed that the increased risk would be minimal and should not be a deciding factor in whether or not to conduct trials with pregnant women.

Question 9: Litigation environment

Do you think litigation is higher in the clinical trial environment or in the postmarketing environment?

> We're risk averse … to anything that has to do with a potential lawsuit.
> —Biotech physician

The participants I interviewed expressed concerns about the potential for litigation against the pharmaceutical companies and how it could impact research and product availability. Most respondents thought that the risk of litigation was lower in the clinical trial environment than the postmarketing environment, followed closely by those who answered, "I don't know."

Only three participants thought that the risk of being sued was lower in the postmarketing environment. The concept of the "learned intermediary" was mentioned by two: "in the post-marketing environment the prescribing physician has the decision-making responsibility," and, for marketed products, "you're going to have a labeled statement about use in pregnancy … and the prescribing physician will have made the judgment about that in light of the known risks." Another of the three thought the risks were higher during clinical trials because you know less about the safety of the drug at that point in time. He felt that the drug being studied could be associated with spontaneous abortions or birth defects that occurred during the trial by chance. The last of the three stated that, "as soon as the Company is involved, automatically you assume that there is a greater risk," but, he added, "it's a guess."

Most of the interview participants thought that the risk of litigation was higher in the postmarketing environment for the reasons shown in Tables 7.1 and 7.2. They felt somewhat protected by the assumption that people participating in clinical trials were aware that the drug was experimental. When the drug is on the market, they felt that the public assumption is that the drug has been shown to be safe and effective and it is used by a larger and more diverse group of people. These factors could increase the risk for adverse pregnancy outcomes that could result in litigation.

TABLE 7.1 Factors perceived to increase the risk of litigation by environment.

Factors that increase the risk in the postmarketing environment	Factors that increase the risk in clinical trials
No informed consent	Little knowledge about the safety of the drug being studied
Lack of adequate testing/due diligence in clinical trials prior to marketing	No regulatory statement about safety (i.e., the drug label)
Many more pregnant women will be taking the drug in an uncontrolled, uninformed manner. They may have concurrent medical conditions or be taking concomitant medications; have less instruction on the proper use of the product and less monitoring for safety and efficacy	No learned intermediary prescribing the product
	Current standard of care is exclusion—why were they testing pregnant women?

TABLE 7.2 Factors perceived to decrease the risk of litigation by environment.

Factors that decrease the risk in the postmarketing environment	Factors that decrease the risk in clinical trials
Learned intermediary (prescribing physician)	Informed consent
Drug label	Drug is known to be experimental
Drug was approved by FDA	Study conducted according to regulations
	IRB acknowledgment that risks were minimized and study design is ethical
	Study is conducted for the benefit of pregnant women
	Legal scrutiny of the protocol prior to implementation
	Select population in trials
	Historical precedent—it is harder to succeed with litigation in the clinical trial setting than in the postmarketing setting

FDA, US Food and Drug Administration; *IRB*, Institutional Review Board.

I actually think the litigation risk would be higher in the post-marketing environment. The clinical trials are being conducted according to regulations, being reviewed by an IRB, people are going into the study being informed about potential risks, and people are in the trials being conducted for the benefit of the specific people or ... pregnant women. It'll be much more difficult to make a case [for] the mother or the fetus who was harmed in the clinical trial setting.

Now if you take that in the post-marketing setting, where you have this drug that's been approved by the FDA and now it has some deleterious effect on the pregnant woman or it's not effective, everyone's going to come down and say, 'how, FDA, could you let this be approved?' and also, 'how, Sponsor, can you allow this to go to the market? You didn't do your due diligence, you didn't do your research to see if it would affect pregnant women, to see if it would be safe.'

I think now you have much more firepower to say you didn't do everything you should have, you didn't do due diligence, you breached your duty, therefore we can make a good case against you. Versus in the clinical trial setting, everyone is aware, everyone knows it's supposed to be experimental. I really think there'll be more chance for litigation in the post-marketing setting.

—IRB member

Litigation risk seems to be higher in the post-marketing stage, because of the fact that in clinical trials, you have a very select population. You have a smaller population in order to get a trial, in order to get a drug approved. And [when] they're using the drug in post-marketing and it goes into widespread use there—many, many, many more patients—and patients who don't necessarily, they are real world patients, they don't fit the profile of a select population for a clinical trial. They may have comorbidities, it's not controlled, it's not under a proscribed set of instructions as to how to take the drug. So you have much more risk. The risk goes up because the proportion of patients taking the drug increases. —Pharma physician

The informed consent document and process were mentioned by many participants as protecting the companies against allegations of research-related injury.

I would think, not being a lawyer, if in fact the consent forms were designed properly for clinical trials, and if the woman had a real opportunity to talk about the pros and cons of the disease, of the drug, and the possible outcomes, I would think litigation in the clinical trials might actually be less than in the post-marketing environment. Because in post-marketing, many people don't get the true, broad benefit/risk analysis of the drug before they start taking it. —Pharma physician

Historically,... it's hard for a plaintiff to succeed if there was informed consent. —Pharma attorney

In spite of this, pharma company stakeholders confirmed a real concern about the potential for litigation. They disclosed that litigation can result in high costs to the company, and damage to a company's reputation.

Obviously, pregnant women and children or babies are hot button emotional topics for juries and so it's not just, 'what is the risk of being sued?' but if you lose, 'how much is the risk for damages?' —Pharma attorney

Squeamish is too benign a term. Apoplectic is more like it. —Pharma physician

The mood is moving strongly in the direction that any abnormality of the child is potentially suspect as a result of poor care by the obstetrician/gynecologist or the result of material that they were exposed to. A new chemical entity or an unregistered chemical entity would be an easy target. —Pharma physician

Liability—this is an emotional, sensitive subject. I can see how in a lawsuit, any harm to a mom or a fetus could play well to a jury. There would be unknown damages, speaking objectively. Therefore, I would caution any sponsor in enrolling pregnant women especially in the absence of data that says it is safe or if it may not be effective—the potential harm would give us pause.

—Pharma attorney

In a similar vein, one of the participants referred to another pharmaceutical company that has a strong reputation for conducting pregnancy registries for their products that are intended for use by women of childbearing potential. For one such product, the company had identified an increased risk for a certain birth defect. Subsequent to this finding, the company was the subject of television and internet advertisements encouraging women who had used the product to call the law firms in the ads. Her concern was that the lesson learned was that a company might be at higher risk for having done a study and found a correlation than if they had not conducted the study at all.

Another cautioned that a little knowledge can be a dangerous thing. It may not be unusual to find a random birth defect in a small sample of pregnant women. "This," he says, could be "suggested as [having] prior knowledge" and could be used against the company in litigation.

There are just not enough pregnant women who are exposed until the medications are on the market. Some of these things can't be studied and can't be evaluated until they are on the market and then you are dealing with ... a less controlled, and more real-world environment.   —Pharma member

Whether or not the risk of litigation is higher in the clinical or the postmarketing arenas, the fear of such litigation is real and may have other consequences. One respondent described its impact saying, "It's very tough. I can tell you that within major pharma, there are drugs that can be very useful and that address a very clear unmet medical need that are being given thumbs down by senior management because of the specter of endless litigation."

In the end, the advice of a company attorney was that, "I'm not sure either way that the litigation issues ought to drive you either to do or not to do trials [in pregnant women]." Stated another, "Well, you never know if you'll lessen the litigation risk, but you know, we accept litigation as the risk of doing business."

Key findings for Question 9

My perception was that this question had not been widely considered by the study participants. But the issue is raised in the literature: pregnant women are using marketed medication that has not been studied in pregnant women.

The result of not testing the products on pregnant women in the controlled clinical trial environment is that pregnant women take medication in the postmarketing environment usually without the benefit of informed consent, risk minimization considerations, and the enhanced monitoring of her pregnancy and the fetus that would be available in a study. Consider the difference in the number of birth defects that occurred when thalidomide was on the market ($>10,000$) compared to the number of defects that might have occurred had it been tested in a clinical trial. A higher number of birth defects in the postmarketing environment might result in a higher risk for litigation.

A few of the respondents considered the risk to be higher in clinical trials because there is less known about the compound being tested. But most presumed it would be higher in an untested drug postapproval. The discussion was speculative, however, because we lack actual experience with the inclusion of pregnant women in clinical trials.

1. Pharmaceutical companies are concerned about litigation risks associated with testing products on pregnant women in both the clinical trial and the postmarketing environment.
2. Fear of litigation may be deterring pharmaceutical companies from testing drugs in pregnant women in clinical trials.
3. Fear of litigation about birth defects may be deterring the development of potential pharmaceutical interventions that address unmet medical needs of the population.
4. There is a fear that evaluating the safety of a drug in pregnant women may increase a company's risk for litigation.
5. The risk of litigation is considered to be higher in the postmarketing environment than in the clinical trial setting.

External support

Question 10: Regulatory agency support

What would it take to have companies open (or IRBs approve) relevant clinical trials to pregnant women? Would a guidance document be strong enough or would it need to be by regulation?

The FDA interviewee thought the proposed changes to the pregnancy section of the label, which have been in development for over 10 years now, might be helpful.

The Pregnancy and Lactation labeling rule, [released in 2010] will have a greater emphasis on human data when it's available and hopefully a better mechanism to get that human data into the label. [H]aving that rule out, that regulation out, might have an effect on this issue because [having] a better mechanism to get it into the label, it may make people more willing to capture that data in a systemized fashion.   —FDA member

Having established that people within industry and related organizations see the need for improved knowledge on how to treat medically compromised pregnancies, the next step was to ask for their input on how to do that. The participants in this study were very experienced in their respective areas and so would likely represent current thinking on the topic and/or could provide suggestions based on experience within their companies and organizations.

I think it would take a woman CEO. People who have had issues with (pregnancy), people who have wanted information and have had to make difficult decisions with the pregnancy are more conscious of these issues than people who haven't. —Biotech physician

I think it would require FDA to have a strong position. And then I think you'd need patient groups that would be pushing. And then I think you would need enlightened researchers in the company that are willing to take the next step for research in the 21st century. I think we're still very far from it.
—Biotech physician

A pharma industry lawyer reasoned that, because there is no regulatory impediment to the inclusion of pregnant women, there is no need for a guidance or regulation. However, most of the interviewees thought a guidance document would be an effective tool to get the dialogue started, to get stakeholders to take notice of the issue, to raise consciousness.

[A guidance] would be your first step to actually having sponsors not be fearful to include pregnant women in clinical trials. And it would be a huge step for the IRB. It is the FDA standing up and saying, 'we support this.'
—IRB member

Respondents considered a guidance to be "a favorable fact in litigation," and "a sanction for enrollment." Without a guidance document, most thought that very little would happen.

...if they're not providing guidance, believe me, IRBs aren't going to want to touch it. —IRB member

While a guidance document may be required before even considering the enrollment of pregnant women, many felt that the guidance alone would not be enough for companies to change their practice of excluding them from the drug development process.

Certainly a good, thoughtful guidance document would be helpful for the really altruistic company or one where this is the nuts and bolts of their indication to treat non-pregnant related illnesses that occur during pregnancy. But my guess is, unless told to do so, most companies would not. —Pharma physician

The experience with studies in children suggests that a regulation would be necessary. —Pharma physician

If we want a universal way of doing it, then I think there needs to be a regulation. Otherwise it will depend on the goodwill and the interests of companies and will be very uneven. —Biotech physician

But not all interviewees were convinced that a regulation is the answer either.

From a litigation perspective, it would be a good defense. But I can't see them saying you have to do it—it would pose a risk for the FDA. —Pharma attorney

Interviewees expressed hope that a regulation would not be required.

I personally have this philosophy of, 'don't give me a rule if I don't need a rule.' Or a law. And while I applaud the success that the pediatric laws have had driving people to the right space, I would just love to think that we could get this just by the force of public need without having to think about regulation. Goodness knows we've got enough of them as it is. —FDA member

I also realize that these large business enterprises called pharmaceutical companies have so much going on that sometimes they don't pay attention unless there's a rule. I'd hate to think we have to go there. I really would love to see this take root without having to go much beyond guidance. —FDA member

Others agreed that guidance documents, while nonbinding, are difficult for companies to ignore, and, for the most part, "companies conform." With guidance documents, said a PhRMA lawyer, "you get additional clarity that is quick and adaptable, easier. Regs [regulations] are too vague, guidance can be more detailed." Another physician observed that guidance recommendations "can be achieved more easily and harmonized more easily" across institutions, states, and even across countries.

Question 11: Incentives

Would patent extensions, like those implemented for pediatric trials, be a viable enticement?

Many participants agreed patent extensions were a viable partial solution:

We pharmaceutical companies love patent extensions, because it takes a lot to get a drug on the market. I think it may be required, because if you're going to take the risk of doing it, the patent extension may make it worth your while.

—Pharma physician

Patent extensions have worked for pediatric exclusivity; it could possibly work in this particular case.   —Pharma physician

A company is taking extra risks that have monetary value.   —Pharma attorney

"Transferable extensions" was a suggestion offered by one physician.

You could either extend the patents or you could have a certificate that allows you to transfer it to another product. So, there the statute says that [if] the manufacturer is developing a drug for a rare and, I think maybe, neglected disease drug and they get it approved, they can transfer the patent extension to another drug. So if you've got a multibillion-dollar drug and you are allowed to get an extension on that drug by developing a new orphan drug, that's a huge incentive. So, [either] extend the patent for the product for which it's developed or transfer the extension to another product.

—Biotech physician

However, others expressed dissatisfaction with their patent extension experience in the pediatric sector.

By the time you complete the pediatric program, get through all of the hoops and things, you still might turn out to be too late and you've lost patent already or they've taken so long that a patent extension doesn't add much. Or with the generic challenges to patents that come up so frequently, the patent extension may not be worth a hoot and holler ... [I]t worked out in one of our cases, that we got the six-month patent extension followed one month later by a patent suit and the judge ruled in favor of the challenger.

—PhRMA member

The FDA participant advised caution.

Patent extensions would be tightly linked to the expectation that we have a rule or a law. So if you go that route it means that you're conceding that we need some kind of regulation. Be careful what you wish for.   —FDA member

There may also be a downside to patent extension in the public sector. "Patent extensions are kind of unpopular among the general public these days," said an IRB member.

The last thing that industry would want is to seem like they're doing this from a profit motive as opposed to a public health concern and certainly they've taken a fair amount of criticism for even the pediatric extensions despite a clearer benefit from a public health perspective. It's hard to think that would be that helpful.   —IRB member

A lawyer at PhRMA also agreed, stating that the current fiscal environment is at odds with the provision of additional exclusivity.

Congress is likely not to grant any more patent extension approaches. I think there's a feeling now, with policymakers for some time now, that it hasn't been a great solution. . . .there have been some perceived cases that . . . were seen as industry trying to get the extra market exclusivity . . . you will get some backlash.   —PhRMA attorney

While patent extensions may not be popular with the public or politicians, they are very effective. More pediatric drug studies were conducted in the 5 postextension years than in the previous 30 preextension years combined.[2] Those studies resulted in drug administration information specific to children in more than 400 product labels.[3]

The parallels to pregnant women are striking: a population that keeps gaining and losing members, physiologies that differ from "normal" adults, reluctance to use pharmacotherapies to treat or prevent illness, questions of consent, and a long history of under- and overdosing and tragic consequences following the use of medications that were not tested in that population. So, too, could be the results: an increase the number of pediatric expert advisors in pharma companies, IRBs, and regulatory agencies; an increase in the number of pediatric pharmacology research units, an increase in the number of drugs and biologics that have been studied in children, and an increase in the number of products with pediatric drug dosing instructions in their labels.[4]

Question 12: Indemnification

Because the perceived risk of company liability was high among industry and IRB participants, I asked the stakeholders, "Would company indemnification be necessary? Is that a realistic option?"

Most respondents did not think that company indemnification was a realistic option.

I think there are instances where clearly things were not done properly and then indemnification doesn't matter to me anymore. Indemnification would not [persuade] me one way or the other. I'm not sure it really works in the final analysis because if you're not doing things properly, you're going to be sued, I don't care what the indemnification says.   —Pharma physician

But other participants, once prompted to think further about the possibility, voiced interest in its potential. An IRB representative was aware of current efforts by a governmental committee to explore this issue further.

There are people who are pushing for national funds to reimburse research injury. The Presidential Commission just recommended that in a recent report, following up on the Guatemala issue. Recommendations were: improved accountability and expanded treatment and support for research subjects injured in the course of [a study], because subjects harmed in the course of research should not bear the cost.   —IRB physician

The "Guatemala issue" refers to a study conducted in 1946–48 and supported by the US Public Health Service that involved the intentional exposure of thousands of Guatemalan citizens to sexually transmitted diseases without their consent.[5] President Obama asked The President's Commission for the Study of Bioethical Issues to investigate that study and ensure that current research standards protected national as well as international research participants.[6]

In a separate project, The President's Commission recommended a study of the current environment to evaluate the need for a national system of compensation that would satisfy the ethical obligation of care for research-related injuries incurred in federally funded studies.[7]

The IRB physician continued, "they cite the national Vaccine Injury Compensation Program (VICP) as the example here." The VICP was enacted in 1986 because litigation against vaccine manufacturers due to adverse events claims following immunization threatened to cause vaccine shortages and reduce population vaccination rates.[8] The legislation was aimed at ensuring a stable market supply of vaccines, and to provide cost-effective arbitration for vaccine injury claims.[9] Other participants disagreed with a parallel between the VICP and potential indemnification for studies with pregnant women, citing vaccines' more significant public health impact and the absence of a comparable market concern as differentiating factors.

In summary, participants in the study suggested that, from their point of view, increasing the inclusion of pregnant women in clinical studies is such a challenging and controversial undertaking that any and all suggestions for how to make it happen should be on the table for consideration.

Well, so a guidance document is interesting but probably would not be sufficient to overcome the other concerns that companies have. Carrots, like a patent extension, also may not be sufficient to overcome if there are serious litigation risks. So, indemnification might actually be important. So, for a society and a Congress that really wants to foster drug development [in this area], that might be the most effective way to do it. So you give a carrot [a patent extension] and a safety net for a specific list of conditions. This list of conditions should be studied and if there is a bad outcome for a pregnant woman enrolled in one of those studies there is indemnity for the company and a separate fund for recourse for the injured party. That might be good. You could look at vaccines as a model.   —Pharma physician

Key findings for Questions 10, 11, and 12

1. Most of the participants believed that a guidance document from FDA on the topic of including pregnant women in clinical trials will increase awareness and discussion within and outside of the pharmaceutical companies, but that it may not be enough to cause a change in current practices.

2. Patent extensions and transferable extensions should be considered cautiously due to negative industry and public perception.

3. Company indemnification should be included when considering all potential solutions to improving knowledge of pharmaceutical therapy for pregnant women.

Notes

1. Clayton, E. W. (1994). Liability exposure when offspring are injured because of their parents' participation in clinical trials. In A. C. Mastroianni, R. R. Faden, & D. D. Federman (Eds.), *Women and health research: Workshop and commissioned papers*. Washington, DC: National Academies Press.

2. U.S. Food and Drug Administration. *Drug research and children*. (2016). Retrieved from https://www.fda.gov/Drugs/ResourcesForYou/Consumers/ucm143565.htm.

3. Milne, C. (2011). The case for pediatric exclusivity. *BioPharm International, 24*, 12. Retrieved from http://www.biopharminternational.com/case-pediatric-exclusivity.

4. U.S. Food and Drug Administration. *Drug research and children*. (2016). Retrieved from https://www.fda.gov/Drugs/ResourcesForYou/Consumers/ucm143565.htm.

5. Rodriguez, M. A., & Garcia, R. (2013). First, do no harm: The U.S. sexually transmitted disease experiments in Guatemala. *American Journal of Public Health, 103*(12), 2122–2126. https://doi.org/10.2105/AJPH.2013.301520.

6. Presidential Commission for the Study of Bioethical Issues. *Moral science: Protecting participants in human subjects' research*. (2012). Retrieved from https://bioethicsarchive. georgetown.edu/pcsbi/sites/default/files/Moral%20Science%20June%202012.pdf.

7. Presidential Commission for the Study of Bioethical Issues. *Moral science: Protecting participants in human subjects' research*. (2012). Retrieved from https://bioethicsarchive. georgetown.edu/pcsbi/sites/default/files/Moral%20Science%20June%202012.pdf.

8. Health Resources and Services Administration. *National Vaccine Injury Compensation Program*. (2018). Retrieved from https://www.hrsa.gov/vaccine-compensation/index.html.

9. U.S. Legal.com. *National Childhood Vaccine Injury Act law and legal definition*. (n.d.). Retrieved October 24, 2018 from https://definitions.uslegal.com/n/national-childhood-vaccine-injury-act-ncvia/.

Chapter 8

Perspectives from the industry: on ethics

You've got disenfranchised women basically. They're truly disenfranchised.
—IRB member

Research ethics

Pharmaceutical industry stakeholders expressed a number of concerns regarding the ethical issues raised by the subject matter of the interview. Traditional medical ethics—nonmaleficence, justice, and autonomy—were raised, along with a suggestion that perhaps feminist ethics would make a contribution to the debate. They mentioned their struggles with issues raised by the concept of including pregnant women in research including society's uneasy relationship with fetal protection, including the abortion debate, and the difficulties in considering and balancing both maternal and fetal benefits and risks. Some passionately described their feelings about the dichotomy inherent in the pharmaceutical industry's mission. Does the company's responsibility to provide medical products to improve the health of the population supersede or follow the corporation's mandate to, at least, remain solvent or, preferably, generate and increase profit?

Question 13. Ethical challenges

One of the challenges in doing research with pregnant women is addressing the ethical issues it raises. What ethical problems do you think are most challenging or important?

"We can't ethically include pregnant women in clinical research" is a common response to the issue. *I think that we overplay that card* was another physician's response. Several interview participants stated that it is unethical *not* to include pregnant women in clinical research: *Incorporating pregnant women is the right thing to do, We should include them from the ethical aspect* and *This is just the simple humanity....*

I received a wide range of responses to this question. In speaking to the interview participants, I found the IRB members to be the most

Pregnancy and the Pharmaceutical Industry. DOI: https://doi.org/10.1016/B978-0-12-818550-6.00008-8

comfortable discussing ethical principles, as would be expected, followed by company physicians. One IRB member discussed the vocabulary of medical ethics; a company physician felt that the industry does not pay enough attention to ethical principles. In contrast, in response to the question, "what ethical problems are raised...," a company lawyer responded, *Nothing comes to mind.*

The Belmont principle is the lingua franca for day-to-day operations of ethics in human subject research. It's the framework around which the regulations are written for sure, and the guidelines. So I think the specific ethical issues are probably most readily accessible in that language. You can ask questions of whether other ethical perspectives, for example feminist ethics, which I'm not deeply into, but I hear as being largely an ethic that emphasizes care as opposed to what is seen as more coldly rational application of principles and precepts. Feminist ethics might have a take on this but I'm not competent to suggest what that take would be.   —IRB member

Companies don't often think in ethical terms, most of them don't understand ethics or bioethics, so even a well-articulated argument goes over their heads. I don't think there's an easy answer—that's a long-term question of trying to improve the knowledge of ethics and bioethics in the population—in business and in medicine.   —IRB member

Nonmaleficence

I think that 'do no harm' is the biggest thing that researchers and physicians face. Do no harm. No one wants to do any harm, said a company physician. This principle, above all the others, was the one most consistently invoked by the interview participants. However, further exploration of the topic revealed the common understanding that clinical research has inherent risks of causing harm. Stated one participant, *I don't believe ... that we can't do it because of an ethical possibility that we may cause harm in people. Everything we do may cause harm in people.* And once it was established that, of course, no one wants to do anyone any harm, the conversation could progress to discuss potential ways to do research with pregnant women that lessens that risk—as is done for all populations that participate in research.

Autonomy

Of course, discussing ways to reduce risk assumes that pregnant women are given the opportunity to participate. At this point in time, as one IRB member stated, *You've got disenfranchised women basically. They're truly disenfranchised.* Not allowing pregnant women the opportunity to potentially improve their medical condition and potentially contribute to generalizable medical knowledge, violates their autonomy—the first principle of the Nuremberg Code,[1] and the Belmont Report.[2] Essentially, the pharmaceutical company is making the decision not to participate for them. But, counters a company physician, *we make those kinds of decisions all the time* in our inclusion and exclusion criteria. Pregnant women would be no different from elderly patients or patients with renal impairment who are not included either. At which point the question becomes, if patients in other excluded populations will potentially have need of a drug in development, will those studies be done, and if so, when?

Informed consent of the mother

One of the ways that clinical research deals with risk in clinical studies is by ensuring that the benefits and risks known at the time of the study are articulated in the informed consent document and shared with potential participants in a manner in which they can understand the contents. This process allows the potential study subject to practice his or her autonomy by making an informed decision about whether or not to participate. Several interviewees stated that the informed consent process is very important to their belief that the clinical study is being conducted in an ethical manner.

> *My ethical position is very simple. I think as long as the people are adequately informed with informed consent and a good discussion, I personally, am in favor of conducting studies in these populations because they do benefit from them and these are important studies that need to be done.* —Pharma physician

> *I think being able to share with [pregnant women] what you do and do not know about the drug with them [is important]. Because at the time that you do this study, you have to say to them that we don't have any data in pregnant women. That has to be told to them and that's why we're doing the trial, to get this information, and you are being helpful.* —Pharma physician

Another participant explained that we need to have gathered as much information as possible regarding benefits and risks prior to conducting the study so as to *lay out clearly for folks that these studies are being done at a point where we have relative confidence on risks so that there's no perceived undue risk because of unknown data.*

Informed consent of the fetus

An issue that was raised by several interviewees, and especially by the IRB participants, was the question of the pregnant woman giving consent for the fetus.

I think the biggest ethical challenge, one of the main reasons why an IRB or Sponsor wouldn't want to conduct research on pregnant women, is that now you don't just have the woman, you also have the fetus. . . . I think the ethical concern is you could potentially be making a decision for two people. Does the father also need to provide consent in this situation or is that encroaching on the mother's autonomy to make her own decisions? So I think there's going to be a lot of concern and confusion over really whose decision is it to enroll in the research.   —IRB member

Other participants strongly disagreed with this assessment.

I think we have similar situations all the time in clinical trials—for instance, Alzheimer's patients. There's a lot of study being done on Alzheimer's patients because there's really no medications out there for them, so it's usually the caregiver that gives the informed consent, because the patients can't give it. And in my experience with Alzheimer's trials, because there's nothing out there to treat it, the regulatory agencies and IRB's are willing to approve drugs—I mean investigational products that do have significant side effects. So if you look at an Alzheimer's program, in my view they haven't given consent, they're having severe outcomes from the study drug. So I mean, don't we do this all the time already?   —Pharma physician

The FDA participant summarized her frustration with the issue stating, *I think we're going to have to get over and put away the issue of* [fetal] *consent.*

Distributive justice

You've got autonomy, primum non nocere, and distributive justice. The question of autonomy and the question of harm are on one side, on the other side you've got distributive justice.   —Pharma physician

When you're talking about ethical considerations, in the Belmont Report, one of the three principles is justice. You need to have an equal distribution of the benefits and the burdens of research. If you're not testing pregnant women you don't have equal distribution of the benefit of the research.
—Pharma physician

Pregnant women are entitled to be considered in research because it's their right. They're like anybody else. If you're going to exclude a portion of your population from research you have to have a good rationale for it. Like if you're going to conduct a study and you're going to exclude African-Americans, well why? Why would you do that? They're entitled to bear the burdens and the risks as well as the benefit of whatever else is out there. As long as the benefits of participating in the research . . . always outweigh the risks.   —IRB member

Finally, a company physician, with an awareness of the history of the development of clinical research in this country, stated, *I agree with AIDS activists and women who want to know the answer to the question.*

Society, the fetus, and abortion

As discussed in previous chapters, there is an uneasy social consensus on the status of the fetus as a person/patient in its own right versus an entity that is wholly dependent on its pregnant mother. Several participants saw a conflict between the pregnant woman's benefit/risk considerations and the fetus' benefit/risk considerations. *From a pure bioethics standpoint, there is a diversion between the risk to the mother and the risk to the fetus. [There's] no social consensus—that may be the reason for the risk aversion,* stated an industry professional association lawyer. But, he acknowledged, *there is not always a conflict between the two.* Another participant barely acknowledged this in a similar statement, *The question here is also the generic ethical question of weighing benefit/risk to a mother against mostly or almost exclusively risk to a fetus. Almost exclusively because improvement in pregnancy outcome by treating a disease is certainly a benefit to the fetus so you have to ask the question that way.*

In a similar vein, the topic of abortion, though not explicitly raised by the issue itself, was implicitly surmised by some of the interviewees.

> *I don't know, it shouldn't, but I suspect it will raise the issue of pregnancy terminations as well, and how women get counseled on that.*   —FDA member

> *You're going to have the major ethical concern, you're going to have board members on the IRB of various different backgrounds and beliefs and different beliefs on abortion as well. So I think another major ethical concern, depending on the board member, is whether or not it would be appropriate, whether to test something that could be harmful not only to the mother but also to the fetus.*   —IRB member

A third interviewee said, *[P]eople debate over whether the fetus—and about the whole abortion debate about the fetus—I think that's the biggest ethical concern.* In this regard, I think it is important to know that concerns about abortion could influence participants' considerations and decisions in debates about the inclusion of pregnant women in clinical research. Such concerns may need to be addressed directly in proceedings on the topic.

Corporate responsibility

Interview participants were sometimes prompted by me to consider the issue of industry's contributions to the good of society versus contributions to their shareholders. How is the need for improved knowledge about how to treat pregnant women reconciled with the costs and risks to the companies? There were mixed responses to this question, some responding that a company's primary concern is to make money:

> You can ask if there is some kind of an ethical requirement borne by the industry that goes beyond the normal operations of business ethics. I would say strictly speaking, no. That is, I find the notion of the pharmaceutical industry duty to care to be rather elusive to say the least. You can, at times, construe the pharmaceutical industry as a branch of medicine, which it's not.
>
> —IRB member

> I don't think that [corporate responsibility] holds enough weight compared to risk. So I personally would want to see protection for the sponsor and the investigator for enrolling women who are pregnant to offset that perceived issue of fairness. I think fairness—fairness has a role in non-profit government-sponsored research. I think it's a little different in a private sponsor conducting research. Conducting research, we make all sorts of decisions about who we want our initial target population to be, what we're pursuing and not pursuing etc. I'm not sure that argument holds enough weight considering the risk of enrolling pregnant women early on.   —Pharma physician

> I have become very cynical over the years. I personally think pharmaceutical companies are in it 99% for the money and the rest of it is just a show. Okay, great, we've got a drug for X, and yes, it does help a lot of patients, but the best part of it is, 'look at all the profit we made!' And we've made Wall Street happy. Wall Street drives Pharma. I think that Pharma is very well regulated. You have safety physicians, you have governance, strict governance to ensure patient safety. People think we make patients' lives better and that's great and all, but in general, the reason why pharmaceutical companies work is because of the profit, to make profits, as Wall Street drives the capitalistic system. So I think in general, you're not going to get companies to do this unless there's something in it for them or there's some sort of guidelines or guidance saying it's important to do this—then maybe... I know that's a bit harsh. But once people realize that, once people realize that this is a big unmet need in today's society, maybe that will help. But it's difficult.   —Pharma physician

On the other side of this question, many of the interviewees articulated a commitment to providing effective and safe treatment to patients as a high ranking corporate responsibility:

> My personal feeling is that we do [have responsibility to our patients] and that's why you and I and many others are in this business. I guess we like to see the company be successful and pay the stockholders but I think the

opportunities to do studies that have significant public health benefits are really no longer in many other people's hands. I think to some degree the pharmaceutical industry is standing alone in this anymore. Even the National Institutes of Health studies that they do on diabetes and other outcome things, basically we donate the drug. They're basically dependent on the pharmaceutical industries even to support their public health focused studies.

—Biotech physician

I think one thing is clear. We have the ethical responsibility to give the best evidence that we can about the use of our drugs by the populations which may be using them, whether or not that's technically an approved population or not. So if we know that every woman getting a urinary tract infection during pregnancy is using our drug, I think we have an obligation to be real clear about the impact of pregnancy on this drug.   —Biotech physician

We know that our drug is being used in pregnancy but we close one eye to it, we'll take your money, thank you, but we don't want to go there? I think that's our responsibility.   —Biotech physician

To further the discussion, participants sometimes offered reasons for some companies' reluctance to include pregnant women in research and suggestions on how to improve the situation:

Why exclude [someone] from something that benefits them? [Because there's] not a strong enough driving factor to enroll them. The upside is not very clear. You'd have to shift the mindset, the tone at the top of the Company with maybe a regulatory guidance to push endorsement. I am less aware of patient advocacy groups for this population than there are for other disease-related organizations or for the pediatric population. I think it would be hard to change that mindset. An agency guidance would be persuasive—but they'd have to duke it out with the lawyers. In this environment, the bottom line is important.

—Biotech attorney

Others raised the belief that doing what is right is good for the company:

I think that potentially it could give the company a very good name if you do it well. A company that is working well with … the regulatory agency, in terms of the regulations and having things in place. It gives the people the impression that these companies are doing a good job, they have really taken care of the situation—pregnant women and breast-feeding women—that it's a very good impression. That's the company that I trust … because, look at this company, they're doing a pregnancy registry, they are doing the study, they have the know-how. It gives you a little bit more security when you use the drug. Now, obviously people go for the price nowadays, but in your mind, if the company is putting these things in place, you kind of trust the company more. So instead of being a cost, you can see it as a perception you give people who are potentially stakeholders that the company is really doing the right thing in safeguarding or placing safety as a priority and it's a trustable company. The company will get more bottom line and then the shareholders will see that and will invest in the company.　　　　　　　　　　　　—Biotech attorney

Key findings for Question 13

1. Study participants cited ethical principles to both justify and condemn the exclusion of pregnant women from clinical research including nonmaleficence, autonomy, and justice, and suggested that feminist ethics might make a contribution to the topic.

2. Informed consent was considered to be an important issue on two counts:
   a. that a pregnant woman has the opportunity to be given informed consent and that the document is complete, honest, and comprehensible, and;
   b. that the fetus be considered to have an interest in the decision to participate in the study.

3. The issue of including pregnant women in clinical research implicitly raises issues of fetal rights, abortion, and divergent perceptions of the fetus in society.

4. While most participants felt that pharmaceutical companies had a responsibility to provide safety and efficacy information for products that would be used by pregnant women, many also acknowledged that business considerations might decide whether research in this area would be conducted. The attitude of senior management and regulatory agency guidance were recognized as factors that could influence such decisions.

Notes

1. The Nuremberg Code. (1947). *Trials of war criminals before the Nuremberg military tribunals under control council law no. 10* (Vol. 2, pp. 181–182). Washington, DC: U.S. Government Printing Office, 1949. Retrieved from <https://history.nih.gov/research/downloads/nuremberg.pdf>.

2. National Commission for the Protection of Human Subjects of Biomedical and Behavioral Research. (1979). *The Belmont report*. Retrieved from <https://www.hhs.gov/ohrp/regulations-and-policy/belmont-report/read-the-belmont-report/index.html>.

Part III

Uniting the regulators, the industry, and the advocates

Chapter 9

The FDA Guidance, public comment, and affinity with industry stakeholders

April 8, 2018. US Food and Drug Administration (FDA) releases the draft guidance, Pregnant Women: Scientific and Ethical Considerations for Inclusion in Clinical Trials, Guidance for Industry. News headlines include, "FDA Open Door to Pregnant Women in Clinical Trials,"[1] "Pregnancy Not a Bar to Trial Participation,"[2] and "FDA Pushes to Recruit Pregnant Women in Clinical Trials."[3] I'm not sure that FDA is pushing to recruit pregnant patients but otherwise the comments were predominately positive and informed although a few journalists pointed out some areas of concern.

I reviewed the guidance with interest to see where FDA recommendations overlapped with the proposals from the industry stakeholders I had interviewed in 2012. I've summarized the FDA document here (read the complete Guidance Document in Appendix I) and discuss similarities, differences, and criticisms.

Summary of US Food and Drug Administration guidance

FDA qualifies its recommendations in the first paragraph of the 14-page document by calling for the "judicious inclusion of pregnant women" accompanied by "careful attention to potential fetal risk."

In the introduction, FDA states that the draft document's intention is to provide a focus for continued discussions among stakeholders including the FDA, pharmaceutical companies, the academic community, institutional review boards (IRBs), and others involved in conducting clinical trials. Its authors clarify that the document applies to both clinical trials that may enroll pregnant women and clinical trials during which a subject may become pregnant. Its stated main intent is to address research studies for medical indications that occur commonly in women of reproductive potential, both chronic and acute—not just for conditions specific to pregnancy.

Pregnancy and the Pharmaceutical Industry. DOI: https://doi.org/10.1016/B978-0-12-818550-6.00009-X
© 2019 Elsevier Inc. All rights reserved.

Some of FDA's reasons for considering the inclusion of pregnant women in clinical trials include:

- Women need safe and effective treatment during pregnancy.
- Failure to establish the dose/dosing regimen, safety, and efficacy of treatments during pregnancy may compromise the health of women and their fetuses.
- Enrollment of pregnant women in clinical trials may offer the possibility of direct benefit to the woman and/or fetus that is unavailable outside the research setting.
- Development of accessible treatment options for the pregnant population is a significant public health issue.

The FDA considers it ethically justifiable to include pregnant women with a disease or medical condition requiring treatment in clinical trials under the circumstances listed in Table 9.1. Most of these considerations were also voiced by the participants in the pharmaceutical industry stakeholder interviews.

TABLE 9.1 Conditions of inclusion in US Food and Drug Administration (FDA) guidance and stakeholder interviews.

FDA guidance	Stakeholder interviews
• There are ethical and clinical reasons to include pregnant women in clinical drug trials	✓
• Include pregnant women in studies when benefits exceed risks	✓
• Include pregnant women in postmarketing studies	✓
• After adequate nonclinical studies completed	✓
• After safety is established in nonpregnant women and/or other sources	✓
• If safety data are limited and other therapy is available	NM
• If efficacy is established in nonpregnant women and cannot be extrapolated or safety cannot be assessed by other study methods	NM
• Include pregnant women in premarketing studies in certain circumstances	✓
• After adequate nonclinical studies are completed	✓
• If direct benefit to the woman and/or fetus is not otherwise available	NM
• If therapeutic options are limited	NM
• If safety data are robust	NM

(Continued)

TABLE 9.1 (Continued)

FDA guidance	Stakeholder interviews
• Retain women who become pregnant in clinical trial	✓
• After unblinding	NM
• After second informed consent signed	NM
• If disenrolled, follow pregnancy to outcome outside of clinical trial	✓
• Collect PK data in Phase II, III trials and from those retained in ongoing trial	✓
• Include an ethicist in creating drug development planning	✓
• Include perinatal and specialist expertise in data monitoring planning	✓
• Consult FDA reviewer early in drug development planning	✓

NM, Not explicitly mentioned by stakeholders in interviews; *PK*, pharmacokinetic.

The FDA guidance included factors to consider when enrolling pregnant women in clinical studies:

- The gestational timing of exposure to the investigational drug in relation to fetal development,
- The incidence of the disease, the severity of the disease (e.g., whether or not it is life-threatening), and the availability of other therapeutic options and their risks,
- If she has no other viable therapeutic options (e.g., drug resistance, drug intolerance, contraindication, drug allergy) to treat a serious or life-threatening disease or condition.

Other FDA recommendations:

- Pregnant women can be enrolled in clinical trials that involve greater than minimal risk to the fetuses if the trials offer the potential for direct clinical benefit to the enrolled pregnant women and/or their fetuses.
- IRBs must consider including one or more individuals who are knowledgeable about and experienced in working with pregnant subjects.

Kudos, comments, and criticisms to the draft guidance

The intention of the publication of this book was the same as the intention of FDA in the publication of its draft guidance—to foster dialogue among the many stakeholders involved in developing drugs and providing care for pregnant women with chronic or acute illness, not just conditions of pregnant women. The length of time it took FDA to write, review, and get the draft document to publication is a testament to the persistence of the dedicated women and men within the Division of Pediatric and Maternal Health in the Center for Drug Evaluation and Research at FDA. I am sure it took a lot of discussion, patience, and fortitude to bring this document to completion. I can only imagine the amount of education, debate, and compromise it must have taken to convince this crotchety, old, yet highly revered institution to suggest the inclusion of pregnant women in clinical trials. I salute your diligence.

As previously mentioned, there was a 60-day comment period following the publication of the draft guidance in which the public was invited to submit comments and suggestions to www.regulations.gov. Fifty-eight comments have been received although some appear to have been submitted more than once. These comments, if not specifically designated as anonymous, are available in the public domain on that website. Notably few pharmaceutical companies submitted comments. The pharmaceutical and biotechnology industry trade associations, Pharmaceutical Research and Manufacturers of America (PhRMA) and Biotechnology Innovation Organization (BIO), provided remarks, as did the American College of Obstetricians and Gynecologists, the Teratology Society and Organization of Teratology Information Specialists (combined comments), the Council for International Organizations of Medical Sciences (CIOMS), American Academy of Pediatrics, National Center for Health Research, Society for Women's Health Research, etc. More than half of the 58 commentaries were submitted by individuals.

Aside from a few crazy comments like, "Stop this Frankenstein-like practice upon innocent women and babies," my review noted mostly favorable remarks, many suggestions, and a few criticisms. There were requests for clarifications and notes requesting less restrictive guidance in certain circumstances.

Many submissions strongly advised against the inclusion of the health and human services (HHS) requirement #5: "If the research holds out the prospect of direct benefit solely to the fetus then the consent of the pregnant woman and the father is obtained in accord with the informed consent provisions of 45 CFR part 46, subpart A, except that the father's consent need not be obtained if he is unable to consent because of unavailability, incompetence, or temporary incapacity or the pregnancy resulted from rape or incest."[4] Commenters noted that the guidance's continuation of the father's consent requirement is more restrictive than the requirement of only one parent's signature for pediatric research, pediatric medical care, and pregnancy terminations and so it seems unreasonable to apply it here. This restriction insults and interferes with a woman's capacity for autonomous decision-making, at least one commenter added.

The CIOMS 2016 Guideline 19, on Pregnant and Breastfeeding Women as Research Participants, states in bold type, "In no case must the permission of another person replace the requirement of individual informed consent by the pregnant or breastfeeding woman," perhaps in reaction to the HHS guidance.[5] A consolidated commentary from 21 distinguished individuals with expertise in ethics and the inclusion of pregnant women in research stated, "conferring personal veto power on the biological father of the fetus for a research intervention that must go through the pregnant woman's body is ethically problematic."[6]

Even if FDA agrees, however, they are incapable of making changes to HHS rules. The distinguished authors' commentary noted earlier includes the statement, "Two focal features of subpart B, which was last revised in 2001, have been identified as out of step with current thinking: namely, its research-related fetal risk standard for nontherapeutic research, and its approach to dual parent consent. We believe this guidance represents an opportunity for FDA to accelerate needed change in these two areas by marking these two issues as topics of potential future evolution."[7]

The fetal risk standard they referred to was another item that received multiple commentaries. The draft guidance addresses this standard in a couple of places:

1. On page 4, citing subpart B, "The risk to the fetus is caused solely by interventions or procedures that hold out the prospect of direct benefit for the woman or the fetus; or, if there is no such prospect of benefit, the risk to the fetus is not greater than minimal and the purpose of the research is the development of important biomedical knowledge which cannot be obtained by any other means."
2. On page 6, "Pregnant women can be enrolled in clinical trials that involve greater than minimal risk to the fetuses if the trials offer the potential for direct clinical benefit to the enrolled pregnant women and/or their fetuses."

There is concern that pregnant women who would like to participate in clinical research that does not directly benefit them but provides meaningful data to help other pregnant women will not be able to do so if the risk to the fetus is more than minimal. This is problematic for the pregnant population that FDA is trying to help. The document itself encourages pharmacokinetic (PK) testing in Phase II and III clinical trials that provide important dosing information but may not directly benefit each volunteer participating in the study. One of the most compelling reasons people site for participating in research studies is altruism, the knowledge that they will be helping others.[8,9] The CIOMS guidance clearly addresses the issue by stating that, "When the social value of the research for pregnant or breastfeeding women or their fetus or infant is compelling, and the research cannot be conducted in non-pregnant or non-breastfeeding women, a research ethics committee may permit a minor increase above minimal risk."[10]

The term "minimal risk" is limited to risk no higher than that encountered in daily life or in a physical examination. This category of risk is so minimal it is sometimes used to justify a waiver of informed consent—the risk is so low that the proposal does not need to go through a full IRB review. The risks involved in drawing blood are above the minimal risk threshold. There is a "minor increase over minimal" risk category that many commenters thought would be more appropriate to this situation. This risk category would allow women to participate in a clinical trial with the sole purpose of providing PK data to inform dosage levels that can change from trimester to trimester.

Another suggestion that was repeatedly made by commentators was for FDA to address the issue of data evaluation. Unless the study is designed specifically to enroll pregnant women, the number of pregnant women who will be enrolled in, or who will become pregnant while enrolled in, a clinical study will likely be relatively small. The pharma stakeholders interviewed also raised this as an issue. Commenters requested guidance on how to interpret data acquired in such a small subset and if such data should be included in the overall data from nonpregnant subjects. Within the new guidance, FDA refers the reader to other FDA guidances for industry that address research involving pregnant women. I've listed some of the most relevant of these documents in Table 9.2.[11]

TABLE 9.2 US Food and Drug Administration (FDA) pregnancy-related guidance documents for industry.

Date	FDA guidance title	Status
April 2018	Pregnant Women: Scientific and Ethical 2 Considerations for Inclusion in Clinical Trials Guidance for Industry	Draft
June 2015	Pregnancy, Lactation, and Reproductive Potential: Labeling for Human Prescription Drug and Biological Products—Content and Format (Small Entity Compliance Guide)	Final
December 2014	Pregnancy, Lactation, and Reproductive Potential: Labeling for Human Prescription Drug and Biological Products—Content and Format	Final
April 2005	Reviewer Guidance Evaluating the Risks of Drug Exposure in Human Pregnancies	Final
October 2004	Guidance for Industry: Pharmacokinetics in Pregnancy—Study Design, Data Analysis, and Impact on Dosing and Labeling	Draft
August 2002	Guidance for Industry: Establishing Pregnancy Exposure Registries	Final

The FDA draft guidance provides more detailed suggestions regarding the planning and design of clinical trials than my interviews did including how to enroll pregnant minors, re-consenting women who become pregnant during a drug study, when to stop a trial, and other situations addressed in HHS requirements, like pregnancy termination, inducements to participate, and parental consent.

The draft guidance also included a very important explanation of research-related risks. The guidance states, "When an IRB considers whether to approve a protocol involving pregnant women, it should consider only those risks and benefits that may result from the research itself (as distinguished from risks and benefits of therapies that subjects would receive even if not participating in the research. For example, if a subject who was taking a drug is subsequently enrolled in a study evaluating the same drug, the drug risks would not be research related)."

Pregnant women who have been prescribed and are utilizing pharmacotherapy would be excellent candidates for Phase IV clinical studies. And the researchers likely would be more enthusiastic about studying them when the risk to the company has been minimized by its adherence to the draft guidance and the FDA's clear explanation of research-related risk.

One of my favorite recommendations, made by several commenters, suggested that upon the release of the final guidance, instead of having to justify their reason for including pregnant women, the company would have to defend their reason for excluding them from a new drug's development plan. I found the first mention of this advice in Vanessa Merton's "The exclusion of pregnant, pregnable, and once-pregnable people (a.k.a. women) from biomedical research"[12] published in 1993 and referring to the inclusion of women, not pregnant women. I believe Ms. Merton would apply this recommendation to pregnant women as well. I think this is a practical, potentially effective, and inexpensive suggestion.

As shown in Table 9.1, the crafters of the FDA draft guidance document and the pharma industry stakeholders that I interviewed were in agreement about the many of the aspects of research design that would minimize risk to pregnant women who enrolled. This indicates to me that this not rocket science—it is human science and we all have a pretty good idea of how to go about it. My conclusion is that it's not the science that is holding us back. It's the other factors highlighted in the interviews that are keeping pregnant women out of pharmaceutical company-sponsored drug studies.

Notes

1. Frieden, J. (2018, April 9). Pregnancy not a bar to trial participation, FDA says. *MedPage Today*. Retrieved from https://www.medpagetoday.com/obgyn/pregnancy/72229.
2. Institute for Patient Access. (2018). FDA opens door to pregnant women in clinical trials. In: *IFPA's patient advocacy blog*. Retrieved from http://allianceforpatientaccess.org/fda-opens-door-to-pregnant-women-in-clinical-trials/.

3. Knowles, M. (2018). FDA pushes to recruit pregnant women in clinical trials. In *Becker's hospital review*. Retrieved from https://www.beckershospitalreview.com/quality/fda-pushes-to-recruit-pregnant-women-in-clinical-trials.html.

4. U.S. Department of Health and Human Services. (2016). *Code of federal regulations, title 45, part 46, subpart B. U.S. Government Printing Office via GPO Access, 140–143.* Retrieved from https://www.gpo.gov/fdsys/pkg/CFR-2016-title45-vol1/pdf/CFR-2016-title45-vol1-part46.pdf.

5. Council for International Organizations of Medical Sciences. (2016). *CIOMS international ethical guidelines for health-related research involving humans.* Retrieved from https://cioms.ch/wp-content/uploads/2017/01/WEB-CIOMS-EthicalGuidelines.pdf.

6. Comment *F*rom Multiple Signatories Anonymous. (2018). *Comment on the FDA notice: Pregnant women: Scientific and ethical considerations for inclusion in clinical trials; draft guidance.* ID: FDA-2018-D-1201-0038. Retrieved from https://www.regulations.gov/document?D = FDA-2018-D-1201-0038.

7. Comment From Multiple Signatories Anonymous. (2018). *Comment on the FDA notice: Pregnant women: Scientific and ethical considerations for inclusion in clinical trials; draft guidance.* ID: FDA-2018-D-1201-0038. Retrieved from https://www.regulations.gov/document?D = FDA-2018-D-1201-0038.

8. Irani, E., & Richmond, T. S. (2015). Reasons for and reservations about research participation in acutely injured adults. *Journal of Nursing Scholarship, 47*(2), 161–169. https://doi.org/10.1111/jnu.12120.

9. Moorcraft, S. Y., Marriott, C., Peckitt, C., Cunningham, D., Chau, I., Starling, N., ... Rao, S. (2016). Patients' willingness to participate in clinical trials and their views on aspects of cancer research: Results of a prospective patient survey. *Trials, 17*, 17. https://doi.org/10.1186/s13063-015-1105-3.

10. Council for International Organizations of Medical Sciences. (2016). *CIOMS international ethical guidelines for health-related research involving humans.* Retrieved from https://cioms.ch/wp-content/uploads/2017/01/WEB-CIOMS-EthicalGuidelines.pdf.

11. U.S. Food and Drug Administration. (2018). *Pregnancy research initiatives: Enhancing health for mother and child.* Retrieved from https://www.fda.gov/ScienceResearch/SpecialTopics/WomensHealthResearch/ucm256927.htm.

12. Merton, V. (1993). The exclusion of pregnant, pregnable, and once-pregnable people (a.k.a. women) from biomedical research. *American Journal of Law & Medicine, 19*(4), 369–451.

Chapter 10

Proposed actions for FDA and the pharmaceutical industry

Reconciling FDA guidance and industry perspectives

Many similarities existed between the responses of interviewed industry stakeholders and the recommendations in the US Food and Drug Administration (FDA) draft guidance, particularly when discussing the design and conduct of the studies in which pregnant women could enroll. Differences consisted mainly of distinctions of scope with industry stakeholders focusing on the larger societal context within which business functions, which is outside the purview of FDA.

FDA and industry stakeholders agree

Why should pregnant women be included in clinical research?

Multiple reasons for considering the inclusion of pregnant women in clinical trials were described by FDA (Table 10.1) and by industry stakeholders (Table 10.2).

TABLE 10.1 US Food and Drug Administration (FDA) guidance reasons for including pregnant women in clinical trials.

FDA guidance reasons for inclusion
Women need safe and effective treatment during pregnancy
Failure to establish the dose/dosing regimen, safety, and efficacy of treatments during pregnancy may compromise the health of women and their babies
Direct benefit to the woman and/or fetus may be unavailable outside the research setting
Development of treatment options for the pregnant population is a significant public health issue
Extensive physiological changes associated with pregnancy may alter pharmacokinetics and pharmacodynamics and directly affect the safety and efficacy of a drug administered to a pregnant woman

Pregnancy and the Pharmaceutical Industry. DOI: https://doi.org/10.1016/B978-0-12-818550-6.00010-6

TABLE 10.2 Industry Stakeholders' reasons for including pregnant women in clinical trials.

Industry stakeholders' reasons for inclusion	
It's the right thing to do	When the benefit exceeds the risk
There is medical need	To aid the company reputation
To assist health care providers	To inform the product label
To fully evaluate the product's safety profile	To develop medicines that treat the population
To improve insurance coverage for medications	As a competitive advantage
To emulate best practices in other special populations like the elderly and pediatrics	

When should pregnant women participate in clinical research?
When their exclusion cannot be justified by scientific rationale:

- When participation in a study provides therapeutic benefit and the anticipated benefits exceed the anticipated risks.
- When there is medical need to treat a particular pregnant woman or pregnant women in general and there is reliable information from preclinical testing or human experience on the teratogenic and developmental risks of the proposed treatment.
- When there are limited therapeutic options.

Where in drug development should research include pregnant women?

- Clinical environment:
  - Pharmacokinetic (PK) testing in Phase II trials.
  - End of Phase III studies designed to include pregnant women in the general study or designed specifically for pregnant women.
- Postmarketing environment:
  - Phase IV clinical studies designed for pregnant women.

What pregnant women should be included in clinical research?

- Pregnant women in need of treatment (whether for pregnancy-related conditions or illnesses unrelated to the pregnancy) can be enrolled:
  - in studies that potentially provide therapeutic benefit and whose potential benefits exceed the potential risks.
  - in studies with more than minimal risk if potential benefit is only available in the research environment.
- Pregnant women already taking approved medications in the postmarketing environment.

• Women who become pregnant during a clinical trial who desire to remain in the study after individual consideration of risk/benefit and reconsent.

Factors for consideration include the risk to the pregnant woman and her fetus from continuation of therapy, discontinuation of therapy, and the effectiveness and risks of alternative therapies (including risk of fetal exposure to the experimental and the alternative therapy).

Perceived barriers and potential solutions

When stakeholders were engaged in dialogue on this issue, physicians, researchers, lawyers, and executives recognized the unintended adverse consequences of pregnant women's exclusion from clinical research. While understanding the need for change from the current practice of general exclusion, they cautioned that change would be difficult and probably incremental. Via the stakeholder interviews, experts on clinical trial design and conduct identified barriers to inclusion and provided potential solutions to the concerns. These are presented in Table 10.3 and are further discussed below.

TABLE 10.3 Concerns and solutions identified by industry stakeholders.

Key concerns	Key potential solutions
Causing harm to the fetus; scientific concerns	Study design; scientific advances in modeling and animal testing; end of Phase III and postmarketing studies
Litigation	Guidance/regulation; informed consent; indemnification; improved awareness of issue in public domain
Enrollment concerns	PK testing on small numbers; partnerships with OPRUs[a] and obstetrical community; data from multiple sources
Negative impact on initial approval	Postapproval studies
Lack of regulatory agency support, unclear regulations	Final FDA guidance document; harmonization of international guidelines; improved accessibility and communication between FDA and Sponsors
Business concerns	Define market; conduct postapproval studies; devise incentives and protections
Lack of experience and know-how	Collaboration, best practices, innovation, science

FDA, US Food and Drug Administration; *OPRUs*, Obstetric-fetal Pharmacology Research Units; *PK*, Pharmacokinetic.
[a]*See below for description of OPRUs.*

Recommendations

Bringing together suggestions and ideas from the stakeholder interviews, the FDA draft guidance document, the literature, and the commentaries to the draft guidance, I have generated a number of recommendations for FDA, for pharmaceutical and biotechnology companies and companies that support them like contract research organizations and institutional review boards (IRBs), for the industry's professional organizations, PhRMA (Pharmaceutical Research and Manufacturers' Association) and BIO (Biotechnology Innovation Organization), and for pregnant women's advocacy groups.

Proposals for pharmaceutical industry: actions to support the enrollment of pregnant women in clinical research

General considerations

- Pharmaceutical companies have a responsibility to provide efficacy and safety information for products intended for women of childbearing potential and for childbearing women.
- Consideration of the need for drug testing in pregnant women should be part of routine drug development for all new molecular entities.
- Evaluation of all products' effects on pregnancy should continue to be monitored after the drugs are marketed and should be ongoing throughout the lifetime of the product, as they are today.
- Every product does not need to be tested in pregnant women. Consider the company's portfolio of drugs/vaccines/devices. Identifying the conditions and drug classes where the need may be greatest among pregnant women can help set research priorities and allocate resources appropriately.

Pharmaceutical companies should consider:

- Participating in dialogue with stakeholders about increasing the inclusion of pregnant women in clinical studies.
- Designing studies for pregnant women, on an individual product basis, including PK studies, pre- and postapproval safety and efficacy studies, pregnancy registries, and other methodologies to improve the company's contribution to best practices for the treatment of pregnant women with medically compromised pregnancies.
- Establishing a policy for including pregnant women in clinical develop-ment and postmarketing studies, and retaining women who inadvertently become pregnant in clinical studies if the benefits outweigh the risks and the woman requests and consents to continue to participate.
- Documenting the reasoning behind their exclusion if no plans to design clinical studies or retain pregnant women in clinical studies are included in the product development plan.

- Defining the conditions under which they would consider studies for pregnant women, for example, prevalence of the condition in pregnancy, risk of no or delayed treatment, safety and efficacy data available for alternative treatments, etc.

Clinical concerns

- All pregnancies that occur during clinical trials should be followed to outcome regardless of whether or not the subject continues enrollment in the trial.
- With their consent, retain women who become pregnant during clinical trials following an individual benefit/risk assessment. Consider the risk of the exposure versus the benefit of the treatment, the risk of discontinuing treatment, the efficacy and safety, including fetal exposure, of any available alternative treatments.
- Clinical studies can be designed to minimize risk. Build on the experience we have gained from prior studies in pregnant women to plan future studies, for example, studies in HIV + pregnant women to prevent vertical transmission and maternal immunization studies.

Efficacy concerns

- Proper dosing for pregnant women can only be gained by conducting PK testing in pregnant women. Such testing can be done on small numbers of women, and could be done with pregnant women who are already taking the approved medication in postapproval studies.
- Since 2004, Obstetric-fetal Pharmacology Research Units (OPRUs) have been receiving funding from the Eunice Kennedy Shriver National Institute of Child Health and Human Development to conduct pharmacology studies on pregnant women "to enhance understanding of obstetrical pharmacokinetics and pharmacodynamics" and "to improve the safety and effective use of therapeutic drugs in women during pregnancy and lactation."[1] Virtually every organ system is affected during pregnancy resulting in changes to drug metabolism. "Data are lacking on the implications of these changes ... and future research is desperately needed."[2] Sponsors should consider partnering with OPRUs to conduct studies that determine correct dosing for pregnant women.
- When there are plans to enroll pregnant women in a Phase III clinical trial, PK data in pregnant women should be collected during the Phase II clinical trials to guide appropriate dosing in Phase III. In situations where pregnant women are enrolled in Phase III clinical trials for a marketed drug, PK data should be collected as part of the trial as recommended in the FDA draft guidance.
- Women who become pregnant while on the investigational drug and consent to remain on the drug can also consent to PK assessments at steady state to collect data on correct dosing during pregnancy as per the draft guidance.

Safety concerns

- We can improve our knowledge of the safety of medication in pregnancy by innovation in preclinical techniques and analysis, by systematic learning from inadvertent pregnancy exposures during clinical trials, in planned trials for pregnant women, and in postmarketing surveillance, pregnancy registries, and epidemiologic studies.
- Some knowledge of a drug's safety in pregnancy can be obtained from doing careful testing in animals. Acknowledging the need for caution, it is important to note that the majority of drugs are not teratogenic, and almost all drugs that are known to be teratogenic in humans are teratogenic in animals as well (the exception was misoprostol which is teratogenic in humans but one study has found teratogenic effects in rats).[3] Continue supporting advances in drug modeling and animal testing.
- In general, Phase I and II clinical trials should be completed or nearly completed in studies that include females of reproductive potential before enrolling pregnant women in later phase clinical trials (FDA draft guidance).

Company oversight

- Companies should have internal maternal health committees composed of subject matter experts within the company to consult on pregnancy-related issues—in study design and planning, in policy making on inclusion and retention in trials, in postapproval activities and Risk Evaluation and Mitigation Strategies.
- An independent pregnancy-specific data safety monitoring board (DSMB) should provide oversight and decision-making functions for open trials, similar to other organ-specific DSMBs.

Business concerns

- Epidemiologic analyses would be helpful to define the market for pharmaceutical use in pregnant women.
- Epidemiological analyses of the prevalence of the condition targeted for treatment by the new molecular entity, access to treatment, the risk of no or delayed treatment, and safety and efficacy data of alternative treatments will help defend the decision to include or exclude pregnant women from clinical testing.
- To avoid delayed initial approval and access to products with established therapeutic benefit for the general population, consider planning to conduct postapproval studies in pregnant women.

Litigation

- There is no evidence that suggests that designing clinical trials for pregnant women will increase the risk of litigation against the company (though experience is limited).
- There is evidence that discovering teratogenicity postapproval in pharmacotherapies not evaluated during development raises the risk for litigation. "The potential for liability is much greater if efforts are not made to detect fetotoxic effects."[4]
- The introduction of an FDA guidance recommending the inclusion of pregnant women in clinical trials under certain circumstances may increase the risk of litigation if clinical studies meeting the guidance recommendations were not designed and conducted for pregnant women.

Company indemnification

- While many of the study participants were doubtful that indemnification was a real possibility, several of the pharmaceutical company and IRB participants recommended not dismissing the concept of company indemnification outright. They thought that the concept should be included when considering all the potential solutions to improving knowledge of pharmaceutical therapy for pregnant women.
- Pharmaceutical industry concern about the cost and the potential harm to a product and a company's reputation in the case of specious litigation is a legitimate barrier to the implementation of efforts to increase the enrollment of pregnant women in clinical research. Acknowledging the potential difficulties and pitfalls in having to "sell" the concept of indemnification to the public and the politicians, the following four points should be considered:

  1. The President's Commission for the Study of Bioethical Issue's recommendation of a national compensation system, which states that: "Because subjects harmed in the course of human research should not individually bear the costs of care required to treat harms resulting directly from that research, the federal government, through the Office of Science and Technology or the Department of Health and Human Services, should move expeditiously to study the issue of research-related injuries to determine if there is a need for a national system of compensation or treatment for research-related injuries."[5]

  2. One of the conclusions of the 2000 University of Texas Medical Branch conference: pregnant and postpartum women as subjects in clinical trials, held "to address the national problem of underrepresentation of pregnant women in clinical trials" was that "[t]here should be a nationally supported mechanism to protect private sponsors

and industry from excessive or inordinate liability claims and to develop incentives to promote industry-supported research on this population."[6] Other medicolegal experts agree.[7,8,9]

3. The success of the Vaccine Injury Compensation Program (VICP).
4. The potential for industry- and institution-sponsored compensation systems.[10]

Proposals for FDA: actions to support industry's enrollment of pregnant women in clinical research

---

For FDA consideration

If the potential therapeutic benefit to the pregnant woman (and fetus) exceeds the risk of including her in a clinical study, then there is no scientific justification for her exclusion. At what point does the agency consider that the Sponsor has "enough" preclinical and clinical data to perform a benefit/risk assessment? What constitutes adequate data? The establishment of best practices for data collection and evaluation can provide standardization, improved knowledge, and protection from litigation.

Business concerns

- Change from the widespread practice of excluding pregnant women from clinical research studies may require regulatory directives, financial incentives, and legal protections.
- Aside from safety issues, valid business concerns must be recognized and addressed before change can be implemented. These include: the additional time and financial costs with little return on investment, potential delays and threats to product approval, and litigation risks—both financial and reputational.
- Business decisions will influence whether research for pregnant women will be conducted. The attitude of senior management and regulatory agency guidance are recognized as factors that will influence the inclusion/exclusion decision.

Litigation

- Sponsors should not be punished for following best practices to ascertain if a product is teratogenic. The financial and reputational costs of a product that has been inaccurately branded teratogenic, for example, Bendectin,[11,12] can be substantial. Therefore, best practices should be defined and standardized.

Previous statistics obtained from COMAH in the United Kingdom show that 30% of the reports were returned to the plant owner for correction [14]. A similar case occurred in Malaysia when 599 errors/omissions were found in thirty safety reports submitted to Control of Industrial Major Accident Hazards (CIMAH) [15]. The errors must be corrected by a competent individual and undergo an approval process by the Department of Occupational Safety and Health (DOSH). Not only is this process tedious, the plant owner, under the COMAH/CIMAH enforcement, is also required to inform the surrounding community about the hazardous process to be operated and its associated risk and danger. The public's understanding of this regard plays an important role in determining the success of the project, and therefore, communication with the public needs to be prioritized. This obligation implies additional cost to ensure smooth communication with the community and to win the public's trust, if not to risk project cancellation and considerable financial losses to the plant owner.

Through the years several projects were canceled due to public objection, for example, the halts of an $8.9 billion plant expansion project that produced paraxylene in Ningbo [16] and the cancellation of methyl isocyanate (MIC) production in Bayer CropScience at West Virginia, USA facility after $25 million was spent to enhance operational safety [17]. These cases prove that the perception of public community on the operation of hazardous process is very important and improper handling of this issue will cause a large investment to go down the drain. In this work, other than COMAH, the hazardous process is bound to fulfill the requirement for packaging, labeling, and transportation. The additional requirement brings to another potential problem. In the study done by Ta et al. [18] involving 150 participants from industry, 2% of the participants admitted that they never read the information provided by labeling. Around 10% of them confessed that they never read the information provided by Safety Data Sheet (SDS). While the intention of SDS and labeling itself to give information's related to the substance being handled, due to negligence this procedure can be the cause of the accident.

All the earlier-mentioned findings prove that, additional procedures to manage hazardous process can lead to additional cost and can lead to the accident due to human error. Therefore, ISD concept that could eliminate the hazards from the source; it is simple procedure and hence the best choice.

7.6 Conclusion

The application of the ISD concept during the early design stage enables a designer to identify hazards and minimize them through modification of the design. In this chapter, the statistics of human error lead to the accidents in chemical process safety, the basic inherent safety principles in process safety management, and how the implementation of inherent safety can reduce human error during operation has been discussed extensively.

The benefits of inherent safety applications toward compliance of regulations, including evaluation other type of control measures and training required to manage hazardous process has been demonstrated and discussed using case study. It can be thus argued and concluded that the application of ISD during the early design stage, can have a great impact on hazard reduction, which is beneficial for safety and is economically attractive for process operation.

Understanding the human factors involved at the very early stage of design is critical to eliminating or reducing potential catastrophic incidents. By using ISD, the inherent risk can be minimized and subsequently can lead to the reduction of human error and better regulatory compliance. However, we cannot underestimate the importance of operator's training for complex and dynamic environment in chemical process industries.

References

[1] K. Kidam, M. Hurme, Analysis of equipment failures as contributors to chemical process accidents, Process Saf. Environ. Prot. 91 (1−2) (2013) 61−78.

[2] CCPS, Inherently Safer Chemical Processes: A Life Cycle Approach, second ed., John Wiley & Sons & AIChE, New York, 2009.

[3] D.A. Moore, M. Hazzan, M. Rose, D. Heller, D.C. Hendershot, A.M. Dowell, Advances in inherent safety guidance, Process Saf. Prog. 27 (2) (2008) 115−120.

[4] K. Kidam, H.A. Sahak, M.H. Hassim, S.S. Shahlan, M. Hurme, Inherently safer design review and their timing during chemical process development and design, J. Loss Prev. Process Ind. 42 (2016) 47−58.

[5] J.P. Gupta, D.C. Hendershot, M.S. Mannan, The real cost of process safety—a clear case for inherent safety, Process Saf. Environ. Prot. 81 (6) (2003) 406−413.

[6] M.Z. Abidin, R. Rusli, F. Khan, A.M. Shariff, Development of inherent safety benefits index to analyse the impact of inherent safety implementation, Process Saf. Environ. Prot. 117 (2018) 454−472.

[7] M. Zainal Abidin, R. Rusli, A. Buang, A. Mohd Shariff, F.I. Khan, Resolving inherent safety conflict using quantitative and qualitative technique, J. Loss Prev. Process Ind. 44 (2016) 95−111.

[8] P.K. Roy, A. Bhatt, B. Kumar, S. Kaur, C. Rajagopal, Consequence and risk assessment: case study of an ammonia storage facility, Arch. Environ. Sci. 5 (2011) 25−36.

[9] UN Industrial Development Organization, Transportation and storage of ammonia, Fertilizer Manual, Springer, 1998, pp. 196−202.

[10] Wiley-VCH (Ed.), Ammonia, in Ullmann's Encyclopedia of Industrial Chemistry, sixth ed., no. v. 2, John Wiley & Sons, 2003, pp. 671−681.

[11] B. Long, B. Gardner, Ammonia storage-a special case, Guide to Storage Tanks and Equipment, Wiley, 2004.

[12] G.S. Lele, Ammonia storage: selection & safety issues, Chem. Ind. Digest. (2008) 85−89.

- Consider a guidance on how to evaluate whether or not a product has caused a congenital anomaly for use in drug safety evaluation and in litigation.
- Company indemnification should be considered as protection against litigation (see below).

Regulatory

- The guidance document from FDA on the topic of including pregnant women in clinical trials will increase awareness and discussion within and outside of the pharmaceutical companies, but it may not be enough to cause a change in current practices.
- Prioritizing studies for pregnant women by those conditions and drug classes where the need is greatest may facilitate acceptance and target resources to where they are needed the most. Consider constructing such a list.
- The ability for Sponsors to access FDA reviewers to discuss study design options and obtain agency advice was cited as an obstacle to drug development. Access and communication between FDA reviewers and company representatives needs to be substantially improved in order to facilitate the planning and conduct of studies in pregnant women. Without communication, the voluntary conduct of such studies will be negatively impacted.
- International harmonization with Council for International Organizations of Medical Sciences and the International Committee on Harmonization and other international organizations would assist global standardization in the current environment where many pharmaceutical companies and their clinical studies are multinational. Consider working groups to accomplish this task.

Incentives and protections

- Fast track review for New Drug Applications (NDAs) that include plans for studies in pregnancy.
- Financial incentives to offset costs (patent protections, tax incentives, research subsidies, innovate new incentives).
- Patent protections (including transferable extensions), while helpful, should be considered with caution due to poor industry experience with their value in the pediatric population and negative public and political perceptions.
- Would a drug's indication for use in pregnancy, due to the small market, qualify the product for orphan drug status? The number of pregnant women needed treatment for some conditions may be fewer than 200,000.

- Facilitate partnerships with National Institutes of Health, CDC, OPRUs, and others to conduct studies in pregnancy.
- Identify diseases and drug classes that are a priority for drug testing for pregnant women so that resources can be targeted efficiently.
- Provide a list of considerations that would result in the exclusion of pregnant women beyond evidence of teratogenicity in preclinical studies or human exposures, for example, conditions that would never or rarely occur in pregnant women, treatment that could usually be postponed until the conclusion of the pregnancy, etc.
- Develop an efficient process within the agency for individual review of NDAs for including pregnant women (i.e., do not recommend testing in pregnant women solely by indication or drug class).
- Encourage generic companies, where applicable, to participate in and contribute to the costs of research on marketed, off-patent products used by pregnant women.
- Consider participating in discussions of company indemnification; provide insight into how the VICP changed the environment and increased vaccine production; draw parallels between that and indemnification for drug testing in pregnant women.

Proposals for Pharmaceutical Research and Manufacturers' Association (PhRMA) and Biotechnology Innovation Organization (BIO): actions to support industry's enrollment of pregnant women in clinical research

---

For PhRMA/BIO consideration

Consider sponsoring the collection of additional data that would be helpful to industry to inform its implementation of the guidance recommendations including:

- Market analysis—what are the expected financial returns—or lack thereof—for the approved or off-label use of a product during pregnancy? What are the expected costs of conducting additional clinical trials for pregnant women? While financial considerations may not be the deciding factors in the decision to conduct such studies, the associated costs must be factored into the total research costs of a product in development.
- Legal analysis on the risk of increased litigation if pregnant women are:
  - retained in clinical trials in which they inadvertently became pregnant.
  - included in clinical trials designed for testing in pregnancy during development (late Phase III) or in the postmarketing environment.
- Legal opinion on potential protections to prevent litigation in clinical and postmarketing environment, including indemnification or a research-related injury compensation system.

PhRMA and BIO should consider

- Producing a position paper for industry on the inclusion of pregnant women in clinical research summarizing the issues.
- Convening a maternal health working group to consider recommendations for the expansion of the inclusion of pregnant women in clinical research.
- Sponsoring legal and market analyses to inform deliberations on the topic.
- Spearheading discussions to consider a system of indemnification or compensation for medical costs incurred for research-related injuries similar to the VICP.

For FDA and PhRMA consideration

Consider an agency-industry-legal working group to explore the feasibility of indemnification or other compensation system for research-related injury. The adoption of the practice of designing and conducting studies for pregnant women may rest on the outcome of this question. Jury awards for children with birth defects and developmental disabilities—rightly or wrongly attributed to drug exposure—can be severe. Litigation costs can remove safe and effective products from the market (e.g., Bendectin). Improvement in maternal health and positive pregnancy outcomes relies upon accurate knowledge of the safety and efficacy of treatment options during pregnancy. Systematic research is required to obtain this knowledge but litigation risk may prevent it.

Proposals for pregnant women's advocacy groups: actions to support industry's enrollment of pregnant women in clinical research

Advocacy groups could consider

- Reactivating efforts to gather stakeholders and other interested parties to discuss the inclusion of pregnant women in clinical trials and how advocates can support efforts to improve their inclusion by:
  - Increasing general public awareness on the topic via lay and professional journals and social media.
  - Increasing the awareness of the obstetrical community via targeted publications and conference presentations.
  - Increasing the awareness of pregnant women via public health campaigns, lay media and social media sites that they frequent.
  - Demystifying and destigmatizing participation in clinical trials; highlighting the benefits while honestly communicating the risks of participation to the general public.

Future steps

Evaluation of the impact of the FDA guidance

When the FDA guidance is final, will the pharmaceutical industry implement its recommendations? While guidance documents are nonbinding, their recommendations are difficult to ignore and companies usually conform—but will it this time?

There are a number of evaluations that were published following the implementation of the pediatric rule requiring clinical trials for the pediatric population. These include, for example, Improving Pediatric Dosing through Pediatric Initiatives: What We Have Learned;[13] Assessing the Effects of Federal Pediatric Drug Safety Policies;[14] and Economic Return of Clinical Trials Performed under the Pediatric Exclusivity Program,[15] etc. The methodology these authors used is easily transferable to the evaluation of advances in the study of pregnant women. Such measures include evaluating changes to drug labeling that include information specific to use during pregnancy including: pregnancy indications, dosing changes in pregnancy, PK information, new safety information, and information concerning efficacy or lack thereof. Cost savings can be calculated from change to current costs for maternal and neonatal hospitalizations, morbidity from maternal adverse drug events, and maternal illness-induced decreased productivity. However, these calculations would have to wait until drug testing in pregnant women was widely implemented.

Cost calculations could actually get at the public health impact of the guidance, which would be very important to know. I anticipate that the return on investment of this initiative may be difficult to measure in cost savings alone because the market for individual drugs used in medically compromised pregnancies is low and the recipients are geographically dispersed. However, studies on the rate, treatment, and morbidity and mortality of specific conditions occurring in pregnant women over time could be conducted. There are many health economists conducting public health program evaluations, return on investment calculations, and cost benefit analyses at the program level, the population level, and at the national level. Journal articles detailing these strategies and methodologies are plentiful.

I would also refer back to the experts in advocacy and communication who recommended the use of framing issues in social justice terms and the use of anecdotes to convey meaning. Could we ask clinicians in the obstetric field to collect narratives, reports, statements, etc., on how the information has impacted their practice of medicine and the lives of the pregnant women and their babies who participated in clinical trials or benefitted from the use of a new medication or a dose-adjusted therapy in pregnancy? That's where the real impact would be found and its collection and communication will rely upon dedicated health care providers, professional organizations, and patient advocacy groups.

Evaluation of potential negative impact and unintended consequences should also be undertaken. Estimating the cost of conducting the trials in pregnant women, increased litigation due to adverse outcomes, and birth defects or other morbidity attributed to drug exposures in clinical trials would be at the top of the list. The industry would certainly monitor the impact of the guidance on the pharmaceutical company—the costs of conducting the trials and the related infrastructure and administration of them—against any financial returns.

How does one measure the return on investment for corporate responsibility (CR) measures? Customer and employee satisfaction scores are suggested as potential "soft-indicators" of impact on CR scores. These scores are not to be underestimated—pharmaceutical companies compete to get higher ratings on CR indicator scales such as the Corporate Social Responsibility Index, the Access to Medicines Index, and the Human Rights Impact Assessment score. An effort to get a question about the inclusion of pregnant women in clinical studies on a pharmaceutical industry-focused measurement scale would focus attention on the issue. How one achieves that goal would be worthwhile exploring.

New policy implementation and evaluation necessitates the need to monitor and measure both intended and unintended outcomes. Based on the outcomes of these measurements, policy modifications can be made for improvement in both the process and the intended outcomes.

Both industry and FDA will need to monitor and evaluate the impact of the final guidance on the inclusion of pregnant women in clinical research.

The evaluation of the guidance's impact on the inclusion of pregnant women in clinical research would be indicated by the actual increase in proportion of clinical trials that (1) do not exclude pregnant women, (2) those that are designed specifically for enrollment of pregnant women, and (3) by an increase in the number of women who became pregnant during a clinical trial and remained in the study.

The evaluation of impact on treatment of pregnant women would be indicated by an increased proportion of drug labels that include human evidence-based data or evidence-based recommendations for (or against) use in pregnancy. Labeling changes would include information specific to use during pregnancy including: pregnancy indications, dosing changes in pregnancy, PK information, and new safety information. Case reports by, or surveys of, practical experience in obstetrical practice might be another source of information.

The evaluation of increased financial costs to industry—the costs of designing, implementing, and recruiting for the trials, related infrastructure and administration, delays in approvals, experience with litigation, etc.—and the offset of such costs by the impact of any implemented incentives should be calculated.

Rarely occurring adverse effects, including birth defects, may not be identifiable until a large number of people have taken the drug. Postmarketing safety surveillance data compliments clinical research data and needs to continue throughout the life-cycle of all products. Serious adverse events should always be assessed to see if the issue could have been identified earlier if appropriate study parameters had been applied.

Benchmarking/best practices

As the practice of including pregnant women in clinical studies will be new to the research community, benchmarking and sharing of best practices will be vital to the continuing improvement of such clinical research practices. Sharing lessons learned by experience would be facilitated by ongoing participation and monitoring by the proposed PhRMA committee on maternal health or other multicompany working groups, for example, the International Society of Pharmacoepidemiologists, and can be communicated via conferences and medical journals.

Additional recommendations in the stakeholders' words

I asked one final question to the pharmaceutical industry stakeholders, to open the dialogue and allow the participants to contribute any additional thoughts or ideas on the topic.

Question 14: other solutions or incentives

"Are there other solutions or incentives you can think of?"

The stakeholders responded with additional comments and topics for further discussion. Many concluded their comments with a pessimistic appraisal such as, "... but I don't think that will work." We don't know what will work, but industry stakeholders kindly donated their time and their imagination to this project, articulating their beliefs and their ideas to add to the common knowledge of how best to aid pregnant women who are ill. I have organized some of these ideas in Table 10.4 under targeted headings.

TABLE 10.4 Stakeholders' observations, suggestions, and potential solutions.

Observations and recommendations from stakeholders	Solutions
On the business	
"The company … has to adopt it from a very high level. It's got a come down from the top, that the company understands the problem and is willing to commit the company resources to doing it. It's got to take upper management to require it — I don't think the clinical monitors are going to embrace it"	Establish commitment at the top
"My hope is that the guidance would at least drive the companies back to reviewing their current feelings on this"	Respond to FDA guidance
"I have this vision of, once we get over the hump of the concerns — which we've done before, first with women, then with kids, now with pregnant women — it should just be part of the normal [drug] development scheme"	Make routine part of drug development
"[Companies] have to see that there's good public relations, that it's good for the company, good for the industry, good for the sector."	Public relations opportunity
"They'd have to see more companies doing these trials … they have to see success in these trials and then I think they would be more interested in perhaps doing it. You'd want benchmarking — the Company would want to be in the middle of that bell-shaped curve. Maybe once some Companies start doing it and it doesn't result in negative outcomes, more companies might start doing it. I don't think there are enough upsides. There's more downside risk than upside benefits"	Benchmark practices and successes
"You face the obstacle of businesspeople without that [scientific] knowledge base. If they would look to their scientific colleagues — but they're looking at the business from an entirely different perspective"	Have science drive the decision
"In pediatrics, there's a market; with pregnant women, not very much. So from a commercial perspective, it is completely unappealing for the Company"	Define the market
On litigation	
"I think if they knew the data well enough and knew that the drug could help those women without severe consequences to the company, without risk to the company, I think they wouldn't get in the way, they wouldn't prevent it from being done"	Legal community advisement in assessing real risks and devising protections
"I think if … there was some liability for not treating the woman when you could save her life or spare her from increased morbidity, I think that might persuade them as well"	

(Continued)

TABLE 10.4 (Continued)

Observations and recommendations from stakeholders	Solutions
On FDA	
"FDA needs to have … more pats on the back, it needs to do some things that will, you know, [give them] 'Atta Boys.' Once they start thinking outside the box with some of these things, they're going to be in the same position they've been [in] which is kind of the whipping dog of Congress and every other group that wants to criticize them for stuff"	Strengthen public advocacy for FDA's efforts
"And there should be a bit of a, "if you don't do this," especially if the drug or vaccine has a high likelihood of being used by pregnant women, this may be harsh but "it may jeopardize your indication or authorization." We had something very similar. [FDA] said if you don't do this elderly study, you are not going to get an authorization and now we're doing an elderly study. We never would have done it unless they said that in writing"	Consider sanctions for not doing studies
"Every month the drug's not approved you lose a lot of money, so early approval is really an incentive as well. I think what [companies] really want is rapid processing of their applications so that it gets reviewed and the FDA makes the decision"	Implement fast track review as incentive
"I'm not sure there's enough dialogue that takes place between pharmaceutical companies and regulatory agencies throughout drug development. I think these issues need to be discussed early and they need to get some real strong guidelines from the regulatory agency. These agencies have a lot of power and pharmaceutical companies really have to listen to them to get the drug approved. Part of the problem too, is when you try to make an appointment to see some of the people in the regulatory agency, they don't have the time and they push back. And I think you really have to be able to talk with them so that you know what they're thinking and they know what you're thinking"	Release the final guidance document; Improve access to dialogue between FDA and Sponsors
"I do think however that some sort of financial break has to be taken into consideration. Our pediatric studies are extremely expensive, high risk, and you may or may not gain any financial benefit from it, which is OK in a way, but then a generic company comes along a month later and benefits …"	Provide some financial incentive to sponsors of studies
On advocacy	
"I think it needs to be on the agenda of PhRMA and BIO. It takes having a number of position papers out there, white papers, symposia, soliciting interest from professional groups, ACOG, AAP, and other groups. It really takes a concerted effort so that they can all write supportive statements and documentation. The point is that it gets into the collective consciousness"	Involve professional, industry, and medical groups

(Continued)

TABLE 10.4 (Continued)

Observations and recommendations from stakeholders	Solutions
"It comes back to advocacy. You need a think tank to be behind you in this, you need a Washington think tank. The Second Wave coalition might be the group you need so that they can develop position papers, they can be in contact with the different stakeholders. You have to get stakeholders onboard. And you have to be in this for the long haul, this is not a one-off.... So you really need to be developing those discussions now in order to have any chance of getting on the [next PDUFA] agenda"	Sustained work by central advocacy group; develop position; involve stakeholders
"For policy advocacy, use anecdotes. They work. The anthrax example is a good one"	Use anecdotes for advocacy
"Starting off with the regulation or guidance, if they're having groups of experts or an advisory committee to the federal government; if you have more and more respected academics, clinicians proposing why it's important to include pregnant women in clinical trials; if you get all of those things out there; use professional organizations proposing guidelines, anything like that, that's going to help pave the way. Or maybe even include pregnant women ... the more and more support you have for that, the more and more it can catch on and you'll get an okay to include them for various conditions"	Involve stakeholders: academics, clinicians, pregnant women; professional organizations
"[This is] going to take a whole lot of work on the outside of getting folks together in those spaces where we all gather professionally, that have these debates around what's needed and how the best to do it. Not under the shadow of impending law. So that was my idealist speech. I don't think that will happen but it would be nice"	Bring stakeholders together in public forums
"I think historically pregnant women are classed in ... the category of vulnerable patient populations and to the extent that there's a push to say, we're not vulnerable, in fact, we're patients who need to understand the implications of taking different treatments. It's almost like a cultural change rather than a regulatory change"	Remove the vulnerable population label from pregnant women
"I don't think that you will have a lobby of pregnant women because it's different [than pediatrics]"	Lobby
"Is the pregnant population as compelling as the pediatric population? I think the perception is that the pediatric population is more underserved"	Advocate for pregnant women's needs
"This is where I think the change is going to come — when companies are consistently asked what their position is. And that could be by the agencies, by the IRB's, could be by the public. They're going to have to hear multiple voices, but particularly IRBs and FDA"	Have companies clarify their position publicly

(Continued)

TABLE 10.4 (Continued)

Observations and recommendations from stakeholders	Solutions
"I think what would help if [there was] more talk about this ethic. If you could get patient groups talking about it, get onto TV, and if people, if it became an actual issue that society cared about. I feel as though it isn't at the moment"	Involve patient groups, advocate in public arena
"Is there a quantitative assessment that could be done? If pregnant women were not protected [from anthrax], what was that cost? Compare mortality and health costs with the risk of birth defects"	Quantify the cost of exclusion
On stakeholders	
"Personally I'd like to see partnerships between a group like the NIH, industry and maybe even third-party payers, to supporting this effort"	Build partnerships
"Mitigate the risk by building better relationships and partnerships for a trial"	
"One would hope that the first company that's going to be really brave to come in and say, 'we want to do this,' will come in . . . to speak to NIH or FDA or whoever, and say, 'here's the trial we want to lay out,' and [they] will not [be hearing] of this for the first time. And they'll be willing to work with each other to make it happen"	
"So to some degree I think the financial responsibility of undertaking things which may be largely of public health interest and not necessarily the pharmaceutical interest, but to support public health interest and science, should really be shared more broadly by a wider group of people. Not necessarily government but maybe the generic companies, public health groups, NIH and those sorts of organizations"	Share responsibility for public health

ACOG, American College of Obstetricians and Gynecologists; *BIO*, Biotechnology Innovation Organization; *FDA*, US Food and Drug Administration; *IRB*, institutional review board; *NIH*, National Institutes of Health; *PDUFA*, Prescription Drug User Fee Act; *PhRMA*, Pharmaceutical Research and Manufacturers' Association.

Notes

1. Obstetric and Pediatric Pharmacology and Therapeutics Branch. (2017). *Obstetric-fetal Pharmacology Research Unit (OPRU) network*. Retrieved from <https://www.nichd.nih.gov/research/supported/opru_network>.
2. Constantine, M. M. (2014). Physiologic and pharmacokinetic changes in pregnancy. *Frontiers in Pharmacology, 5*, 65. http://dx.doi.org/10.3389/fphar.2014.00065.
3. Philip, N. M., Shannon, C., & Winikoff, B. (Eds.). (2002). *Misoprostol and teratogenicity: Reviewing the evidence. Report of a meeting at the Population Council, New York, May 22, 2002*. Retrieved from <http://www.misoprostol.org/downloads/Teratogenicity/miso_terato_review.pdf>.

4. Clayton, E. W. (1994). Liability exposure when offspring are injured because of their parents' participation in clinical trials. In A. C. Mastroianni, R. R. Faden, & D. D. Federman (Eds.), (1999). *Women and health research: Workshop and commissioned papers.* Washington, DC: National Academies Press.

5. Presidential Commission for the Study of Bioethical Issues. (2011). *Moral science: Protecting participants in human subjects research.* Washington, DC. Retrieved from <https://bioethicsarchive.georgetown.edu/pcsbi/node/558.html>.

6. Goodrum, L. A., Hankins, G. D. V., Jermain, D., & Chanaud, C. M. (2003). Conference report: Complex clinical, legal, and ethical issues of pregnant and postpartum women as subjects in clinical trials. *Journal of Women's Health, 12*(9), 864. http://dx.doi.org/10.1089/154099903770948087.

7. Resnik, D. B., Parasidis, E., Carroll, K., Evans, J. M., Pike, E. R., & Kissling, G. E. (2014). Research-related injury compensation policies of U.S. research institutions. *IRB: Ethics & Human Research, 36*(1), 12−9.

8. Beh, H. (2005). Compensation for research injuries. *IRB: Ethics & Human Research, 27*(3), 11−15. http://dx.doi.org/10.2307/3564074.

9. Henry, L. M., Larkin, M. E., & Pike, E. R. (2015). Just compensation: A no-fault proposal for research-related injuries. *Journal of Law and the Biosciences, 2*(3), 645−668. http://dx.doi.org/10.1093/jlb/lsv034.

10. Henry, L. M., Larkin, M. E., & Pike, E. R. (2015). Just compensation: A no-fault proposal for research-related injuries. *Journal of Law and the Biosciences, 2*(3), 645−668. http://dx.doi.org/10.1093/jlb/lsv034.

11. Brent, R. L. (1995). Bendectin: Review of the medical literature of a comprehensively studied human nonteratogen and the most prevalent tortogen-litigen. *Reproductive Toxicology Review, 9*(4), 337−349. http://dx.doi.org/10.1016/0890-6238(95)00020-B.

12. Brody, J. (1983, June 19). Shadow of doubt wipes out Bendectin. *New York Times.*

13. Rodriguez, W., Selen, A., Avant, D., Chaurasia, C., Crescenzi, T., Gieser, G., . . . Uppoor, R. S. (2008). Improving pediatric dosing through pediatric initiatives: What we have learned. *Pediatrics, 121*(3), 530−539. http://dx.doi.org/10.1542/peds.2007-1529.

14. Dor, A., Burke, T., & Whittington, R. (2007, June). Assessing the effects of federal pediatric drug safety policies. *The George Washington University Medical Center Newsletter,* 1−16.

15. Li, J., Eisenstein, E. L., Grabowski, H. G., Reid, E. D., Mangum, B., Schulman, K. A., . . . Benjamin, D. K. Jr. (2007). Economic return of clinical trials performed under the Pediatric Exclusivity Program. *Journal of the American Medical Association, 297*(5), 480−488.

Chapter 11

A chance at change

Kingdon's workstreams theory

In 1995, John Kingdon proposed a "Policy Window" theory of change in his book, Agendas, alternatives, and public policies.[1] In it, he identified three relatively independent issue workstreams whose interactions are required to advance social change. Kingdon called the participants in the workstreams "policy entrepreneurs," people who are "willing to invest their resources in return for future policies they favor."

The three issue "streams," problem (recognition), policy (proposals), and political (influence), can move along independently until a point in time when they "converge," often due to external forces. This convergence allows the issue and its potential solutions to be recognized across parties. The "window of opportunity," if capitalized on by the entrepreneurs, can put the issue on the political agenda for resolution by the parties involved. The result is the advancement of social policy.

Entrepreneurs in the problem recognition stream identify, describe, and frame an issue as a problem when it may not have been recognized as such before. Problem definitions often have an emotional values component which helps them to get on the agenda for change.

Entrepreneurs in the policy stream contribute potential solutions to a "primeval soup" in which "ideas confront, compete, and combine with each other" and eventually result in policy formulation.[2] The process relies on groups of interested and knowledgeable parties to propose multiple solutions that are both "technically feasible and consistent with policymaker and public values."[3] These policy entrepreneurs "must possess knowledge, time, relationships, and good reputations."[4]

The political stream is critical to getting the issue on the agenda for solution. The policy entrepreneur "recognizes the problem, attaches an appropriate policy proposal to it, and floats the policy proposal in various forums"[5] to bring it to the attention of the people with the power to place it on the agenda for change. Political events may occur unrelated to the issue at hand. Astute policy entrepreneurs can recognize the relationships among the event, the problem, and its proposed solutions and connect the streams. The result of the convergence of two or three of the streams is that

Pregnancy and the Pharmaceutical Industry. DOI: https://doi.org/10.1016/B978-0-12-818550-6.00011-8

"a compelling problem is linked to a plausible solution that meets the test of political feasibility."[6]

Kingdon's Policy Window theory of change is most useful when "capacity exists to act on policy windows."[7] He recommended special studies of the social issue to (1) provide indicators of the existence and magnitude of the issue and to (2) promote constituent feedback.

Advocates, policies, and politicians

A brief timeline may be helpful to appreciate the work that has been going on to support the inclusion of pregnant women in clinical research. This is not a new endeavor. Women and men, public and private advocacy groups, and government committees have identified and promoted and written and met and recommended inclusion for the past several decades. The momentum for actual progress may be building.

In 1992, subsequent to the actions of AIDS and women's health activists, National Institutes of Health (NIH) Office of Research on Women's Health, mandated that the Institute of Medicine form a committee to study the ethical and legal issues pertaining to the exclusion of women from clinical studies. There was a growing realization that the research community's singular focus on, and inclusion of, men in clinical trials denied women the advances being made in medical diagnosis and therapy. NIH requested that the committee (1) consider the ethical and legal implications of including pregnant women and women of childbearing potential in clinical studies; (2) provide practical advice for consideration by NIH, institutional review boards (IRBs), and clinical investigators; and (3) examine known instances of litigation regarding injuries to research subjects and describe issues of legal liability and possible protections.[8]

In 2000, a conference was convened by the University of Texas Medical Branch to "address the national problem of underrepresentation of pregnant women in clinical trials."[9] A multidisciplinary faculty and attendees developed consensus statements on the clinical, ethical, and legal environment to encourage awareness and debate.

In 2003, the Obstetric-fetal Pharmacology Research Unit Network was founded through the US NIH to identify, characterize, and study drugs of therapeutic value in normal and abnormal pregnancies.[10]

In 2007, the Second Wave Consortium held an invitation-only workshop at the American Society for Bioethics and Humanities annual meeting to formulate an ethical framework with which to approach the issue.[11]

In a 2010 conference on maternal and pediatric drug safety, a speaker suggested that it was time for the industry to be brought into dialogue on this issue. The role and perceptions of industry had been identified as a gap in knowledge among advocates. The perceptions of the pharmaceutical industry about the barriers to and opportunities for a broader inclusion of pregnant women in clinical research needed to be ascertained.

In 2012, this project estimated that pregnant women were not included in 95% of Phase 4 clinical trials. Interviews found that pharmaceutical company employees, IRB members, and PhRMA (Pharmaceutical Research and Manufacturers' Association) stakeholders were mostly unaware that the exclusion of pregnant women from clinical research was perceived to be a problem.

In April 2016, US Senator Susan Collins, along with six bipartisan cosponsors, introduced Senate bill 2745, Advancing the NIH Strategic Planning and Representation in Medical Research Act to the Senate floor. The plan addresses research, training, the biomedical workforce, and collaboration with other agencies to meet goals. Section 8 states, "HHS must establish the Task Force on Research Specific to Pregnant Women and Lactating Women to report on issues including development of safe and effective therapies for such women."

In 2017, the 21st Century Cures Act was signed into law by President Barack Obama and established the Task Force on Research Specific to Pregnant Women and Lactating Women (PRGLAC). The Task Force was charged with "providing advice and guidance to the Secretary of Health and Human Services (HHS) on activities related to identifying and addressing gaps in knowledge and research on safe and effective therapies for pregnant women and lactating women, including the development of such therapies and the collaboration on and coordination of such activities."[12]

In April 2018, the US Food and Drug Administration (FDA) released its Draft Guidance for Industry: Pregnant Women: Scientific and Ethical Considerations for Inclusion in Clinical Trials. Public comment on its suggested recommendations was requested and a final guidance will be forthcoming.

In September 2018, a task force mandated by the 21st Century Cures Act presented the Secretary of HHS with a report on the gaps in knowledge and research on the safety and efficacy of pharmacotherapy for pregnant and lactating women (see more on this in Chapter 12: After the guidance). The Secretary will decide if the US government should recommend changes in current practices to improve care for these two groups of women.

The problem stream

Kingdon found that "problems ... are matters of interpretation and social definition"[13] and that issues are only perceived to be problems when there is pressure to do something about them.[14] It was clear from my interviews that no one in industry was working on a solution to this problem because they had not perceived it to be a problem in the first place. Nor did they feel any pressure from external stakeholders—health-care providers, pregnant women, professional groups, or support organizations—to address the issue. *There's just been no interest in looking at this ...* stated one interview participant; *Who's really advocated for clinical trials in pregnant women?* The stakeholders suggested that the lack of interest from within and outside of the companies provided little incentive to initiate change.

However, during the course of the interviews, all of the participants expressed an understanding of the potential problems associated with the exclusion of pregnant women from research studies and they were able to suggest potential solutions. I found that the anthrax/amoxicillin PK study provided a "persuasive and compelling" problem illustration and it was all that was needed to shift the perception from exclusion being a normal and ethical practice to it being a practice in need of reevaluation in light of its potential harms. These key informants did not think that change would be easy (one referred to the *sea change that we have to have here,* another suggested that the general public is *apoplectic* about issues involving the fetus), but they did see the need for further thought, if not action, on the topic. These results identified the need for broader problem recognition among the stakeholders.

Very few interview participants had heard of the Second Wave Consortium. Likewise, very few were aware that FDA had a guidance document (Responsible Inclusion of Pregnant Women in Clinical Trials) on the docket for development. Obviously, this lack of awareness precluded work on a solution.

The project also found that the industry is perceived to be a powerful decision-maker in control of the inclusion and exclusion parameters of a clinical trial. While the FDA, the IRBs, and the institutions at which the research is conducted were all perceived to have "veto power"—that is, they could reject a study protocol proposed by the industry—participants agreed that they had little power to demand that pregnant women be included if the company was not in favor. Therefore targeted advocacy to industry will be important to change the current status.

These findings suggest that improved awareness of the problem is needed both within and outside of the pharmaceutical industry. The work of advocacy groups is recognized as being essential to the effort to broaden the inclusion of pregnant women in clinical research.

The release of the draft FDA guidance document will provide the impetus to move from little internal discussion of the issue to dialogue across the industry. Once the document was released, the industry needed to respond. The release of the document opened the policy window.

I was advised by one of the participants that, *It takes having a number of position papers out there, white papers, symposia, soliciting interest from professional groups, ACOG, AAP, and other groups. It really takes a concerted effort ... The point is that it gets into the collective consciousness. At the moment there is nothing.* He continued, *And you have to be in this for the long haul...*

Kingdon agreed. "Of all the attributes of successful policy entrepreneurs that I could name, sheer persistence is probably the most important."[15]

When I tell people that the topic of my book is the inclusion of pregnant women in clinical trials, the usual and immediate response is, "we can't do that." People associate pregnant women and medication with thalidomide. I have found it to be a challenge to explain the issue in a way that allows them to see the benefits as well as the risks of participation in research. In addition, protection of the fetus is a prominent and sometimes volatile issue in current US society. Even the University of Texas Medical Branch 2000 conference on the issue stated that "the topic is controversial" and limited the participants to invited guests because of what they termed, the "perceived sensitive nature of the topic."[16]

The American Public Health Association Legislative Advocacy Handbook states that public health advocates should "use data and the public health human interest stories that you encounter in your workplace to further your advocacy efforts."[17] One key informant advised me, *For policy advocacy, use anecdotes. They work. The anthrax example is a good one.* Public health advocates have learned to use experience and imagery to communicate facts in a way that captures the audience's imagination and helps the messages stick in the recipients' minds. In "framing an issue," advocates "select some aspects of a perceived reality and make them more salient in a communicating text, in such a way as to promote a particular problem definition, causal interpretation, moral evaluation, and/or treatment recommendation for the item described."[18] Communicating about and supporting the position of increasing the number of pregnant women in clinical research can be a challenge.

To address that challenge, I found the anthrax treatment example to be helpful in conveying the seriousness of the lack of evidence-based treatment guidelines. In the examples below, bioterrorism, death, and abortion provide context for the consequences of underrepresentation of pregnant women in research. The use of such frames can assist advocacy efforts during periods of policy change.

- At the time of the anthrax scare in 2002, the recommended treatment for anthrax-exposed pregnant women was 500 mg amoxicillin three times a day for 60 days. Subsequent study, published in 2007, revealed that this dosage and frequency would be ineffective against anthrax due to the effects of pregnancy on the pharmacokinetics of amoxicillin. The 2007 study recommended "further research ... to determine appropriate antibiotic regimens for pregnant women in response to a bioterrorism attack."[19]

- In my work on a pharmaceutical company pregnancy registry, an obstetrician reported that a pregnant woman, concerned about the effect that drugs might have on her developing fetus, made the decision to stop using her asthma medications. When the patient was 7-month pregnant, she experienced an acute asthmatic episode and died.
- Pregnant women in the pregnancy registries have been advised by their doctors to terminate their pregnancies due to their inadvertent exposure to drugs or vaccines. None of the products were suspected to cause birth defects. Whether the advice was given to protect the physician or the pregnant women was not determined.
- Pregnant women were sicker and died at a higher rate than nonpregnant women in the 2009 H1N1 flu epidemic.[20] In response, the manufacturer of Tamiflu performed pharmacokinetic analyses and found that blood concentrations were lower and renal clearance was higher in the pregnant women. They suggest that higher dosing or more frequent administration may be needed for Tamiflu to be effective for women with flu who are pregnant or recently delivered.[21]

In addition to using anecdotes to illustrate issues, Dorfman et al., in their paper on Framing Public Health Advocacy to Change Corporate Practices,[22] recommend articulating "core messages that correspond to shared values." They cite Daniel Beauchamp's 1976 recommendation to frame issues using the public health core value of social justice.[23] The values of social justice include shared responsibility, interconnection and cooperation, strong obligation to the collective good, assurance of basic benefits, government involvement, and community superseding individual well-being. I think these values may particularly resonate with individuals in the pharmaceutical industry, many of whom have a background in medicine, nursing, pharmacy, and basic sciences—disciplines that promote the discovery and application of interventions that improve the public health. I am less confident about our current social and political affiliation with these ideals.

Examples of messaging using core values of public health and social justice:

- It was the tragic outcomes from the use of thalidomide by pregnant women that triggered the 1962 FDA amendments that require efficacy and safety information be obtained from clinical trials prior to drug approval. It is especially ironic then, that pregnant women remain systematically excluded from the benefits of inclusion in clinical trials.
- If clinical studies for thalidomide had included pregnant women, the teratogenicity of the drug would have been identified way before more than 10,000 babies around the world were born with, or died from, severe deformities.

- Pregnant women are the last of the "research orphans." It is likely that more children, elderly, and inmates benefit from inclusion in clinical trials than pregnant women.
- American women are dying of pregnancy-related complications more than women in any other developed country. The United States is the only such country in which the rate is rising.[24]
- For every 100,000 live births in 2014, 12 white American women died. For every 100,000 live births in 2014, 40 black American women died.[25]

In addition to clinicians and scientists, the pharmaceutical industry is also composed of business people. "The biggest barrier to achieving social justice," state Dorfman et al., "is the competing ethic of market justice."[26] The business concerns of the industry were cited by the stakeholders in this study as powerful justifications to continuing to exclude pregnant women from participation in clinical trials.

Former US Surgeon General Antonia Novello said that "one of the fundamental paradoxes of market-oriented societies is that some entrepreneurs—even acting completely within the prescribed rules of business practice—will come into conflict with public health goals."[27] That reflects the issue here; there is no mandate or requirement to broaden the inclusion of pregnant women—it is simply the right thing to do. Framing the problem from a public health perspective (showing the values behind the reason for change) and using anecdotes to illustrate the issue at a fundamental level (making it personal) may help move the dialogue toward collective solutions rather than entrenched positions. Language, word choice, and symbols are important and can be used to promote selected interpretations, mobilize support, and influence the political environment.[28]

Leadership in the problem stream

In "Building the Next Generation of Leaders," Joy Phumaphi, former Botswana Minister of Health says, "A leader who tries to drive the health agenda alone lacks vision. Every stakeholder needs to feel a part of the solution. To reach this point, all must see the problem."[29]

Martin McKee, professor of public health at the London School of Hygiene and Tropical Medicine, states, "Effective public health leaders should not simply wait to be asked for their opinion. They should be advocates for health, drawing attention to issues that would otherwise be overlooked." He continues that, because "public health is based on social justice ... its advocates will often espouse causes that are unpopular."[30] I can confirm that advocating for the inclusion of pregnant women in clinical studies is not a popular stance to take within a pharmaceutical company. I am also a public health professional with experiential knowledge of the issue as a clinician who has treated pregnant women and as a researcher who has gathered data in pregnancy registries to inform treatment decision-making for pregnant women for many years. My position, which was shared by some pharma stakeholders, is that industry can "evolve" in the manner in which it considers and addresses the needs of pregnant women and their health-care providers.

Advocacy calls upon nontraditional leadership models to promote change. Advocates often use "Transformational Leadership" skills, leading with passion, inspiration, and relationships[31] rather than authoritative leadership practices that are derived from positions of power. Advocacy leaders rely upon strong interpersonal skills to explore issues and communication skills to add stakeholder voices to the advocacy and problem-solving process.[32] The authors cited in the bibliography, who have eloquently described the plight of the pregnant woman who is ill, are many of those advocacy leaders who are ensuring that "all will see the problem."

The policy stream

The central "policy" here is the draft FDA Guidance, Pregnant Women: Scientific and Ethical Considerations for Inclusion in Clinical Trials. Although guidance documents, according to FDA, "are not legally binding, they show stakeholders one way to reach their regulatory goal. However, stakeholders are free to use other approaches that satisfy the relevant law and regulations."[33]

Release of the FDA draft guidance has and will continue to instigate the pharmaceutical companies to consider the inclusion of pregnant women in drug development planning. Changes to the current process will ultimately be reflected in internal company policies and procedures. This is where change will really happen.

Kingdon's Policy Window theory has been described as "an evolutionary model of public policy."[34] In the policy formulation stream, Kingdon says that, "Specialists try out and revise their ideas by ... attending conferences, circulating papers, holding hearings, presenting testimony, writing reports, publishing articles, and drafting legislative proposals." The resulting "primeval soup" of policy proposals then go through "the process of policy evolution, [where] some ideas fall away, others survive and prosper, and some are selected to become serious contenders for adoption."[35]

Kingdon's "window of opportunity," if capitalized on by the entrepreneurs, puts the problem on the political agenda for resolution. This "agenda" can be a government agenda (key topics on the policy development list) or a decision agenda (developing policies that are moving into position for a definitive decision).[36] We are dealing with both.

During the period of public comment, interested stakeholders submit their perspectives, preferences, and proposals for consideration. In the policy stream, the formation and refining of policy proposals is a process by which ideas confront, compete, and combine with each other, forming combinations and recombinations.[37] Stakeholders, all of whom may respond to the call for public comment, include the general public, pregnant women, their families and their health-care providers, women's health advocates, the obstetrical community, maternal/child health organizations, IRBs, the pharmaceutical industry, etc.

Kingdon recommends that the development of proposals be done before the opportunity to submit them arises. One of this project's stakeholders from PhRMA characterized the industry association as being reactive, not proactive. He was, indeed, unaware that the FDA guidance on this topic was on the docket—information that was available in the public domain. To me, the pharmaceutical industry seemed unprepared for the new FDA guidance and may have missed an opportunity to influence its content.

Should pregnant women be included in the drug development process? The inclusion of pregnant women in clinical research has serious practical and financial implications for the pharmaceutical industry. It has the potential to change the way companies perform their core business activity—the development of pharmacotherapy to prevent, ameliorate, or cure disease. Each pharmaceutical and biotechnology company will need to consider the implications in the context of their own enterprise. Current internal policies and procedures about the drug development process will change; new policies will be implemented. The results of my investigation are intended to assist the companies and their trade associations, PhRMA and BIO (Biotechnology Innovation Organization), to consider the implications and formulate their responses to the document—to add their voice to the public debate that was going on without their participation.

Guidance documents can be viewed as a type of regulatory policy. Although they do not mandate behavior, their contents are closely adhered to by the industry. As they explain within many guidance documents themselves, "In general, FDA's guidance documents do not establish legally enforceable responsibilities. Instead, guidances ... should be viewed only as recommendations, unless specific regulatory or statutory requirements are cited. The use of the word 'should' in Agency guidances means that something is suggested or recommended, but not required."[38] FDA regulations, on the other hand, "limit the discretion of individuals and agencies, or otherwise compel certain types of behavior."[39] Noncompliance with voluntary adoption of guidance documents can result in mandatory regulations. Pharmaceutical companies want to avoid regulations, so adoption of the recommendations contained in guidance documents is the common approach.

Leadership in the policy stream

"Public health leaders must contribute to national debates; problems that governments face in relation to public health are difficult, and they cannot expect to solve them on their own. Public health leaders ... contribute to solving these problems. The most successful public health leaders have engaged in the policy process."[40]

In 1983, a rural sociologist, Everett Rogers, published his theory of Diffusion of Innovations[41]—how change is adopted by individuals and organizations. In it, he called "diffusion" the process by which a new idea or a change in thinking (an innovation) is communicated by members of a social system. The diffusion process includes persons becoming knowledgeable about an issue, persuading others to make a change, making decisions based on the new thinking, implementing the innovation, and then via evaluation, either confirming it as worthwhile or discarding it as ineffective. This process takes place over time. The final FDA guidance document will spur the adoption of the innovation and the industry is certain to evaluate its utility and its cost.

The "policy cycle" is a familiar construct in public health and is applied to this issue in Table 11.1.[42] Knowledge of the cycle assists entrepreneurs, advocates, and others to understand problem-solving and where in the cycle to intervene to influence policy.

The political stream

"[T]here is ... broad agreement that politics and political issues are rarely analyzed and frequently ignored at all stages of the policy identification, development, and implementation process in the health sector."[43]

TABLE 11.1 The policy cycle.

Stages in policy cycle	Phases of problem-solving	Description and comments
Agenda setting	Problem recognition	FDA creates and releases draft guidance to public
Policy formulation	Proposal of solution	Stakeholders formulate and submit perspectives and policy options
Decision-making	Choice of solution	FDA considers all comments to create final guidance
Policy implementation	Putting solution into effect	Stakeholders consider options, formulate internal policies
Policy evaluation	Monitoring results	Stakeholders monitor implementation, intended and unintended consequences, and continue dialogue with FDA

FDA, US Food and Drug Administration.

FDA put the issue in the political stream when it released its draft guid-ance document, Pregnant Women in Clinical Trials: Scientific and Ethical Considerations, in April 2018. The government agenda was made clear—regulatory guidance for the pharmaceutical industry to increase the inclusion of pregnant women in clinical studies.

Some participants in the policy stream come to the convergence more prepared than others. FDA and the Second Wave Consortium have been working on this issue, often behind closed doors, for many years. The pharmaceutical industry, in my mind, comes to this issue in the weakest position of the three major players. Unlike its power position in determining whether to include pregnant women in individual clinical trials, it appears to be unprepared for the debate about the inclusion of pregnant women in clini-cal research in general.

Nonetheless, industry is a powerful political constituent with well-organized lobbyists and a network of connections. It is also the subject matter expert on how to develop drug therapies and design clinical trials to test their effectiveness and safety. While the industry may be in a weakened position because it has not been paying attention to the issue, it will be moti-vated to respond when the time comes. As Thomas Oliver points out, "Any proposed change to policy threatens the existing distribution of benefits and costs, and groups with an identifiable stake in the outcome," like the pharma-ceutical industry, "will organize themselves in the political system" in response. He continues, "The targets of regulatory policies can make policy implementation extremely difficult. Organizations ... facing concentrated costs will likely continue to resist or seek opportunities to renegotiate the original policy."[44]

Leadership in the political stream

"Public health professionals who understand the political dimensions of health policy can conduct more realistic research and evaluation, better antic-ipate opportunities as well as constraints on governmental action, and design more effective policies and programs."[45] The inclusion of pregnant women in clinical studies has obvious—and difficult—political implications. From the involvement of a powerful and highly regulated industry to the publicly debated (or privately held) opinions regarding fetal protection and women's reproductive health, the topic of what's best for pregnant women is often controversial and politically charged.

"Public policy is not a single act of government but a course of action that involves individuals and institutions in both the public and private sectors, and encompasses both voluntary activities and legal injunctions."[46] Currently, pharmaceutical companies can make voluntary and individual decisions as to the inclusion of pregnant women in the clinical trial process. Sometimes, when the industry has been slow to adopt initiatives voluntarily, regulations will follow—as was seen with the pediatric drug testing initiatives. With the release of the final guidance on the inclusion of pregnant women in clinical trials, the industry will have to weigh the pros and cons of adopting new innovations in the assessment of drug safety and efficacy in pregnant women. Other stakeholders will monitor whether those efforts are enough to improve the evidence base for treatment decisions—or if legislative action will be needed.

One of the stakeholders in the interviews, who has significant political insight and experience, encouraged pursuing this issue via the legislative process:

> *Politically, I think it needs to be on the agenda of PhRMA and BIO. It's got to get on their agenda so that the industry or the sector as a whole can be supportive to varying degrees. This is the kind of thing that can show up in the next PDUFA (Prescription Drug User Fee Act) renewal ... there's a fair lead time. So you really need to be developing those discussions now in order to have any chance of getting on [the next PDUFA] agenda ...*

PDUFA, first passed in 2002 and renewed every 5 years since, allows FDA to collect user fees from the companies who are applying for a new drug approval to help pay for the resources required to perform the application's review. In response to complaints of prolonged reviews restricting access to new treatments, especially for HIV, the Act significantly improved the number of reviewers at FDA and decreased the amount of time it takes for a drug—or device or biologic—to get through the approval process. FDA, PhRMA, and the general public were all in favor of the regulation. In 2007, the PDUFA renewal was part of the Food and Drug Administration Amendments Act which included initiatives like requiring pediatric drug testing, rewarding developers of treatments for neglected diseases, and mandating the posting of clinical research studies on ClinicalTrials.gov. The stakeholder was suggesting that continuing advocacy, research, and leadership by advocates on this issue (or no voluntary response from the industry) could lead to regulation that will provide incentives, encouragement, or requirements to include pregnant women in clinical research in the next PDUFA renewal in 2022. Pharmaceutical companies participate in this process through PhRMA or BIO representation.

Whether legislation will be needed or whether the voluntary approach will provide enough positive change to be acceptable to the stakeholders will be decided by the impact of the guidance publication, public advocacy, and the government (HHS) initiative via the PRGLAC taskforce. Adoption of innovation by the pharmaceutical industry may be the determining factor as to whether legislative reform will be required. Additional research, monitoring, and evaluation will be needed. And ongoing leadership via advocacy, communication, persuasion, bargaining, positional power, and political pressure will be required.

The participation of pregnant women in clinical research has advocacy, policy, and political implications. In order to elicit change in social issues, Kingdon recommended obtaining constituent feedback along with indicators of the existence and magnitude of the issue via research, which was the motivation for this project. Even with the release of the final FDA guidance, I believe much effort will need to be sustained over many years to make the inclusion of pregnant women in clinical research a reality.

Notes

1. Kingdon, J. (1995). *Agendas, alternatives, and public policies* (2nd ed.). New York: Harper Collins College.
2. Lieberman, J. M. (2002). Three streams and four policy entrepreneurs converge: A policy window opens. *Education and Urban Society*, *34*(4), 445. https://doi.org/10.1177/00124502034004003.
3. Stachowiak, S. (2009). *Pathways for change: 6 theories about how policy change happens.* Organizational Research Services. Retrieved from <http://nmd.bg/wp-content/uploads/2014/04/TW1_Pathways_for_change_6_theories_about_how_policy_change_happens.pdf>.
4. Stachowiak, S. (2009). *Pathways for change: 6 theories about how policy change happens.* Organizational Research Services. Retrieved from <http://nmd.bg/wp-content/uploads/2014/04/TW1_Pathways_for_change_6_theories_about_how_policy_change_happens.pdf>.
5. Lieberman, J. M. (2002). Three streams and four policy entrepreneurs converge: A policy window opens. *Education and Urban Society*, *34*(4), 445. https://doi.org/10.1177/00124502034004003.
6. Brieger, W. R. (2006). *Policy making and advocacy. [Powerpoint® slides].* Retrieved from <http://ocw.jhsph.edu/courses/SocialBehavioralFoundations/PDFs/Lecture19.pdf>.
7. Stachowiak, S. (2009). *Pathways for change: 6 theories about how policy change happens.* Organizational Research Services. Retrieved from <http://nmd.bg/wp-content/uploads/2014/04/TW1_Pathways_for_change_6_theories_about_how_policy_change_happens.pdf>.
8. Mastroianni, A. C., Faden, R., & Federman, D. (Eds.). (1994). *Women and health research: Ethical and legal issues of including women in clinical studies.* Washington, DC: National Academy Press; Institute of Medicine.
9. Goodrum, L. A., Hankins, G. D. V., Jermain, D., & Chanaud, C. M. (2003). Conference report: Complex clinical, legal, and ethical issues of pregnant and postpartum women as subjects in clinical trials. *Journal of Women's Health*, *12*(9), 864. https://doi.org/10.1089/154099903770948087.
10. Zajicek, A., & Giacoia, G. P. (2007). Clinical pharmacology: Coming of age. *Clinical Pharmacology and Therapeutics*, *81*(4), 481−482. https://doi.org/10.1038/sj.clpt.6100136.
11. Lyerly, A. D., Faden, R. R., Harris, L., & Little, M. O. (2007). *The second wave: A moral framework for clinical research with pregnant women.* Retrieved from <http://asbh.confex.com/asbh/2007/techprogram/P6364.HTM>.
12. *NIH Task Force on Research Specific to Pregnant Women and Lactating Women. Report to Secretary, Health and Human Services Congress.* (2018). Retrieved from <https://www.nichd.nih.gov/sites/default/files/2018-09/PRGLAC_Report.pdf>.
13. Rochefort, D. A., & Cobb, R. W. (1993). Problem definition, agenda access, and policy choice. *Policy Studies Journal*, *21*(1), 57. https://doi.org/10.1111/j.1541-0072.1993.tb01453.x.
14. Redington, L. (2009). *The Orphan Drug Act of 1983: A case study of issue framing and the failure to effect policy change from 1990−1994* (Unpublished doctoral dissertation). Chapel Hill, NC: University of North Carolina. Retrieved from <https://cdr.lib.unc.edu/search/uuid:a012aad2-1ab1-43b2-b5ab-0e14740e5e07?anywhere = Redington>.

15. Kingdon, J. (2005). The reality of public policy making. Chapter 6. In M. Danis, C. Clancy, & H. R. Churchill (Eds.), *Ethical dimensions of health policy*. New York: Oxford University Press.
16. Goodrum, L. A., Hankins, G. D. V., Jermain, D., & Chanaud, C. M. (2003). Conference report: Complex clinical, legal, and ethical issues of pregnant and postpartum women as subjects in clinical trials. *Journal of Women's Health, 12*(9), 864. https://doi.org/10.1089/154099903770948087.
17. American Public Health Association. (2005). *APHA legislative advocacy handbook: A guide for effective public health advocacy*. Washington, DC: American Public Health Association.
18. Entman, R. M. (1993). Framing: Toward clarification of a fractured paradigm. *Journal of Communication, 43*(4), 52. https://doi.org/10.1111/j.1460-2466.1993.tb01304.x.
19. Andrew, M. A., Easterling, T. R., Carr, D. B., Shen, D., Buchanan, M., Rutherford, T., et al. (2007). Amoxicillin pharmacokinetics in pregnant women: Modeling and simulations of dosage strategies. *Clinical Pharmacology & Therapeutics, 81*(4), 547−556. https://doi.org/10.1038/sj.clpt.6100136.
20. Louie, J. K., Acosta, M., Jamieson, D. J., Honein, M. A.; for the California Pandemic (H1N1) Working Group. (2010). Severe 2009 H1N1 influenza in pregnant and postpartum women in California. *New England Journal of Medicine, 362*(1), 27−35. https://doi.org/10.1056/NEJMoa0910444.
21. Beigi, R. H., Han, K., Venkataramanan, R., Hankins, G. D., Clark, S., Herbert, M. F., et al. (2011). Pharmacokinetics of ostemavir among pregnant and nonpregnant women. *American Journal of Obstetrics & Gynecology, 204*(6 Suppl. 1), S84−S88. https://doi.org/10.1016/j.ajog.2011.03.002.
22. Dorfman, L., Wallack, L., & Woodruff, K. (2005). More than a message: Framing public health advocacy to change corporate practices. *Health Education and Behavior, 32*(3), 320−336. https://doi.org/10.1177/1090198105275046.
23. Beauchamp, D. E. (1976). Public health as social justice. *Inquiry, 13*, 101−109.
24. Martin, N., & Montagne, R. (2017). *U.S. has the worst rate of maternal deaths in the developed world*. National Public Radio/WHYY. Retrieved from <https://www.npr.org/2017/05/12/528098789/u-s-has-the-worst-rate-of-maternal-deaths-in-the-developed-world> citing Global Burden of Disease 2015 Maternal Mortality Collaborators. (2016). Global regional, and national levels of maternal mortality, 1990−2015: A systematic analysis for the Global Burden of Disease Study 2015. *The Lancet, 388*(10053), 1775−1812. https://doi.org/10.1016/S0140-6736(16)31470-2.
25. *Centers for Disease Control and Prevention. Reproductive health: Pregnancy mortality surveillancesSystem*. (2018). Retrieved from <https://www.cdc.gov/reproductivehealth/maternalinfanthealth/pregnancy-mortality-surveillance-system.htm>.
26. Dorfman, L., Wallack, L., & Woodruff, K. (2005). More than a message: Framing public health advocacy to change corporate practices. *Health Education and Behavior, 32*(3), 320−336. https://doi.org/10.1177/1090198105275046.
27. Novello, A. C. (1992). Smoking and health in the Americas. Atlanta, GA: U.S. Department of Health and Human Services, Public Health Service, Centers for Disease Control, National Center for Chronic Disease Prevention and Health Promotion, Office on Smoking and Health. DHHS Publication No. (CDC) 92-8419.
28. Redington, L. (2009). *The Orphan Drug Act of 1983: A case study of issue framing and the failure to effect policy change from 1990−1994* (Unpublished doctoral dissertation). Chapel Hill, NC: University of North Carolina. Retrieved from <https://cdr.lib.unc.edu/search/uuid:a012aad2-1ab1-43b2-b5ab-0e14740e5e07?anywhere = Redington>.
29. Phumaphi, J. (2005). Building the next generation of leaders. In W. H. Foege (Ed.), *Global health leadership and management*. San Francisco, CA: Jossey-Bass.
30. McKee, M. (2005). Challenges to health in Eastern Europe and the former Soviet Union: A decade of experience. In W. H. Foege (Ed.), *Global health leadership and management*. San Francisco, CA: Jossey-Bass.

31. London, M. (2008). Leadership and advocacy: Dual roles for corporate social responsibility and social entrepreneurship. *Organizational Dynamics*, *37*(4), 313–326. https://doi.org/10.1016/j.orgdyn.2008.07.003.
32. Fabrizio, C. S. (2011). *Physician's perceptions of the Hong Kong Cervical Screening Program: Implications for improving cervical health* (Unpublished doctoral dissertation). Chapel Hill, NC: University of North Carolina. Retrieved from <https://cdr.lib.unc.edu/record/uuid:831aabd1-e0d2-4dbb-a89c-934a603b618a>.
33. *US Food and Drug Administration. Fact sheet: FDA good guidance practices.* (2017). Retrieved from <https://www.fda.gov/AboutFDA/Transparency/TransparencyInitiative/ucm285282.htm>.
34. John, P. (2003). Is there life after policy streams, advocacy coalitions, and punctuations: Using evolutionary theory to explain policy change. *Policy Studies Journal*, *32*(4), 488. https://doi.org/10.1111/1541-0072.00039.
35. Kingdon, J. (2005). The reality of public policy making. Chapter 6. In M. Danis, C. Clancy, & H. R. Churchill (Eds.), *Ethical dimensions of health policy*. New York: Oxford University Press.
36. Lieberman, J. M. (2002). Three streams and four policy entrepreneurs converge: A policy window opens. *Education and Urban Society*, *34*(4), 445. https://doi.org/10.1177/00124502034004003.
37. Kingdon, J. (2005). The reality of public policy making. Chapter 6. In M. Danis, C. Clancy, & H. R. Churchill (Eds.), *Ethical dimensions of health policy*. New York: Oxford University Press.
38. *US Food and Drug Administration. Guidance for industry: Enforcement policy concerning certain prior notice requirements.* (2011). Retrieved from <https://www.fda.gov/RegulatoryInformation/Guidances/ucm261080.htm>.
39. Wikipedia Contributors. (n.d.). Policy. In *Wikipedia, the free encyclopedia*. Retrieved October 26, 2018, from <http://en.wikipedia.org/w/index.php?title = Policy&oldid = 484514172>.
40. McKee, M. (2005). Challenges to health in Easter Europe and the former Soviet Union: A decade of experience. In W. H. Foege (Ed.), *Global health leadership and management*. San Francisco, CA: Jossey-Bass.
41. Rogers, E. M. (1983). *Diffusion of innovations* (3rd ed.). New York: Free Press.
42. Association of Faculties of Medicine of Canada. (n.d.). Primer on public health population: The policy cycle. In M. Howlett, & M. Ramesh (Eds.) (1995), *Studying public policy: Policy cycles and policy subsystems*. Toronto: Oxford University Press. Retrieved from <http://phprimer.afmc.ca/Part3PracticeImprovingHealth/Chapter14DecisionMakingPolicies AndEthicsInHealthCareAndPublicHealth/Thepolicycycle>.
43. Glassman, A., & Buse, K. (2008). Politics and public health policy reform. In K. Hehhenhougen & S. Quah (Eds.), *International encyclopedia of public health* (Vol. 5, pp. 163–170). San Diego, CA: Academic Press.
44. Oliver, T. R. (2006). The politics of public health policy. *Annual Review of Public Health*, *27*, 195–233. https://doi.org/10.1146/annurev.publhealth.25.101802.123126.
45. Oliver, T. R. (2006). The politics of public health policy. *Annual Review of Public Health*, *27*, 195–233. https://doi.org/10.1146/annurev.publhealth.25.101802.123126.
46. Oliver, T. R. (2006). The politics of public health policy. *Annual Review of Public Health*, *27*, 195–233. https://doi.org/10.1146/annurev.publhealth.25.101802.123126.

Chapter 12

After the guidance

Women's health care advocates and organizations have been raising the issue of pregnant women's inclusion in clinical research for many years now. It has been on the agenda for the formulation of US Food and Drug Administration (FDA) guidelines since at least 1993, yet the people I spoke within the pharmaceutical industry were barely aware of it as an issue—like a flea on the back of an elephant, hardly noticeable. Meanwhile advocacy groups were holding meetings to discuss the issue, raise awareness, and decide on a path forward—but they failed to ask representatives from the industry to join them in their deliberations. The pharmaceutical industry was perceived to be the behemoth in the unyielding barrier to the advancement of evidence-based pharmacotherapy for pregnant women.

My point-in-time review of Phase IV studies (the category that would provide the safest protocols in which pregnant women could participate) confirmed the practice of exclusion—95% of these trials would not enroll them (see Chapter 4: A measure of exclusion). In key informant interviews, stakeholders (physicians, lawyers, executives, researchers) from the pharmaceutical and biotechnology industries (including drug and vaccine manufacturers), institutional review boards (IRBs) (ethics committees), Pharmaceutical Research and Manufacturers' Association (PhRMA), and Biotechnology Innovation Organization (BIO) (the industry's professional groups), and related parties identified barriers to, and opportunities for, broadening the inclusion of pregnant women in clinical research. The stakeholder interviews described in this book confirmed that the pharmaceutical industry excludes pregnant women based primarily on the desire to avoid harming a fetus, the economic intent to avoid the financial and reputational risk of potential litigation, and to forestall additional time and costs to the drug development process. When engaged in dialogue on this issue, industry stakeholders recognized the attendant, though unintended, adverse consequences of pregnant women's exclusion from clinical research. Many agreed with the need to modify current practices but were pessimistic that the culture of exclusion would change.

While the 16 people I interviewed are not a representative sample of the industry, I nevertheless felt that the perspective of this small sample, who I

Pregnancy and the Pharmaceutical Industry. DOI: https://doi.org/10.1016/B978-0-12-818550-6.00012-X

felt were very candid in their conversations with me, were enlightening—and preferable to the resounding silence from the industry that existed in the public and professional discourse at the time. Prior to the release of the FDA guidance document, the stakeholders I interviewed from within and around the pharmaceutical industry were able to articulate the problems resulting from pregnant women's exclusion from clinical trials and make specific recommendations for research practices that were very similar to those recommended by the long-awaited draft FDA guidance released in 2018. They identified important potential barriers from the perspective of the industry that is largely responsible for pregnant women's participation and liable for its attendant legal risks. The implementation of the guidance's recommendations will need the recognition and management of these barriers if the advocacy is to succeed. The release of the FDA guidance puts the industry on notice that the time for change has arrived.

A white paper compiled from my research and interviews was sent to FDA in response to the Call for Public Comment after release of the FDA draft guidance in April 2018. Commentary was also submitted to FDA from members of the public, individual pharmaceutical and biotechnology companies, PhRMA and BIO, organizations concerned with healthy mothers and babies, and other stakeholders. FDA will review and respond to these comments and proposals. The final guidance, Pregnant Women: Scientific and Ethical Considerations for Inclusion in Clinical Trials, will be published at an undetermined date in the near future.

The pharmaceutical industry perspective

Experienced researchers within the pharmaceutical companies—experts on clinical trial design and conduct—can readily provide practical solutions in study design to overcome the perceived barriers to pregnant women's inclusion in clinical research studies. Yet pharmaceutical company researchers and lawyers, and colleagues in associated organizations like IRBs and trade associations, think that change to the current practice of excluding pregnant women from clinical research studies will be difficult to implement. They believe that, aside from safety issues, valid business concerns must be recognized and addressed before change will be considered. These issues include: the additional time and financial costs with little return on investment, potential delays and threats to product approval, litigation risks—both financial and reputational, and challenges in data interpretation and validity. Based on these factors, they believe that change to current practice, if initiated, would likely be incremental.

Inducing change in the well-entrenched practice of excluding pregnant women from clinical research studies may require regulatory directives, financial incentives, and legal protections.

According to the industry stakeholders' feedback, the FDA draft guidance, and the Task Force on Research Specific to Pregnant Women and Lactating Women (PRGLAC) recommendations (see later), women with medically compromised pregnancies can receive treatment within research protocols that minimize the risks to the mother and the fetus. The benefits of the research studies must exceed the risks and the pregnant woman must be receiving therapeutic benefit or the protocols would be deemed unethical and would not be approved by an IRB. (Whether it is permissible for pregnant women to participate in research with more than minimal risk for the sake of generating knowledge and not just for therapeutic benefit is under debate. Numerous commentaries raised this issue to FDA for clarification.)

Practices recommended in FDA guidance documents are not always adopted by the pharmaceutical industry. As opposed to a regulation that is mandatory, there is a fair amount of interpretation allowed in a guidance, which has been described as a document that "... does not draw a new line in the sand so much as provide more precise warnings about the location of the existing lines."[1] Based on the results of the interviews with the stakeholders, I remain unsure that the industry will listen to the many arguments in favor of including pregnant women in clinical studies. I think that the primary aversion to inclusion—the desire to do no harm to the fetus—can be overcome by innovations in preclinical testing, research study design, and the timing of participation in the studies as discussed earlier in this book and in the FDA guidance. The economic justifications for the exclusion of pregnant women in clinical research—an increase in research costs and potential delays, and a perceived risk of litigation—may be harder to overcome, especially in the absence of financial reward. The current economic climate is not friendly to potentially costly new initiatives based on considerations of social justice, shared responsibility, cooperation, or obligation to the collective good. Such proposals may be "nice-to-haves" that do not make the cut during a financial analysis. The pharmaceutical industry is currently executing downsizings, mergers, outsourcing, and decreases in research and development budgets in an effort to cut costs and maintain profit in a global economy. So, while the stakeholders I spoke with from the companies may understand the need to include pregnant women in research and individually support their inclusion, they are not necessarily the business leaders who must make the difficult decisions that cut potentially promising programs or decline to support corporate responsibility proposals.

The alternative to broadening the inclusion of pregnant women in clinical studies is to continue the current practice of their exclusion. It is risky, but FDA guidelines can be ignored. The Obstetric-Fetal Pharmacology Research Unit sites will continue to conduct pharmacokinetic (PK) testing on priority medications for important diseases; pregnancy registries will continue to collect exposure outcomes over time; population-based registries in European countries will provide data; and electronic health and insurance records will enable case–control studies—all of which, however, take years to arrive at informative data. In the meantime, pregnant women will continue to be misinformed, mistreated, overdosed, and undertreated. I fear that, unless public pressure is applied, significant incentives are offered, and/or protection from litigation is devised, the status quo may remain the norm. Advocacy efforts and economic pressures from other influential stakeholders may be needed.

John Kingdon theorized that change can occur if the advocates in the problem stream convincingly (and persistently) articulate the issue as a problem (which has been achieved among some professional groups but remains lacking among the public), the FDA and the commentators in the policy stream compose an acceptable solution (currently in progress at the agency), and actors in the political stream (regulatory authorities and, potentially, legislators) provide sufficient pressure to induce change (which remains to be seen). Such a trifecta might then ensure that, when the policy window closes, the new strategy behind the pane will be favorable to pregnant women.

The pregnant woman's perspective

Finally, one has to ask the question, "Will pregnant women participate in drug studies?" The struggle to allow pregnant women to participate in clinical trials will be moot if no one chooses to enroll.

Concern has been raised that it may not be possible to enroll enough pregnant women to achieve statistical significance. Because pregnant women have been routinely excluded from clinical studies we do not know to what extent they might volunteer to participate. Studies have shown that pregnant women tend to overestimate the risk of environmental exposures (including drug exposures),[2,3] which suggests they would be less inclined to participate in clinical research.

Pregnant women could only be invited to participate if they are in need of treatment, if the study would potentially provide therapeutic benefit (or, perhaps, advance scientific knowledge), and if the potential benefits exceeded the potential risks. In this way, their treatment in a research study would be similar to their treatment in clinical practice—with the added benefit of improved informed consent, enhanced pregnancy monitoring, and the knowledge that she has contributed her experience to the accumulated medical knowledge base to assist other pregnant women. Currently, evidence from pregnant women treated in clinical practice is rarely captured at all.

Research confirms that people participate in clinical research studies for many reasons aside from accessing otherwise unavailable treatments. These motivations include: closer medical monitoring than in routine practice, enhanced attention from study providers for other ailments, better physical and laboratory health checks, superior physicians, labs, and testing, more contact with the providers, remuneration, and contributing knowledge and experience to society.[4] Pregnant women would likely participate in Phase IV studies for the same reasons that nonpregnant people would.

One small study found that 95% of pregnant women interviewed said that they would participate "if there is a chance that participation in a clinical trial would help their pregnancy and improve their baby's health."[5] They may even be *more* likely to participate than the general population. Sick pregnant women, while being averse to exposures that might cause harm, are vitally interested in anything that would improve or protect the health of their babies. Wild et al. conducted interviews with 50 pregnant or recently pregnant women.[6] Most were reluctant to participate in research studies even though their health would probably benefit. However, when the scenario changed to one in which the research would be beneficial to the baby, even though it would be risky to her, most of the women said they would agree to participate. The researcher speculated that in the interest of helping her fetus, she may downplay the risks to herself and may even be vulnerable to coercion. Overplaying the benefit to the baby may coerce a woman to take risks she otherwise would not rationally take.

Further research should be done to articulate the voice of this key constituency as patient advocacy will be necessary to achieve change. Patient advocates were an important constituency in changing clinical trial policies to include women of childbearing potential in clinical studies, to include infected women in human immunodeficiency virus studies, and to conduct studies on breast cancer treatment. Obstetricians, midwives, nurse practitioners, childbirth educators, and other advocates for pregnant women need to be aware of the ongoing effort to improve the pharmacotherapeutic evidence that informs clinical practice. Educating the ever-changing pregnant population will be a challenge especially in the current climate of intense anxiety and extreme caution about pregnancy exposures to a wide variety of environmental factors, foods, and medications. Pregnant women who are ill will need to understand the risks and benefits of experimental treatments as well as the risks and benefits of alternative treatments and of nontreatment. Without their participation in research, we will not improve evidence-based care—with or without FDA guidance.

Where are we now?

The US 20th Century Cures Act[7] was enacted in December 2016. Its intent is to accelerate the development of and access to clinical innovation, health care advances, and pharmacotherapy. It also seeks to improve the recruitment and retention of experts in science, medicine, and technology, modernize research infrastructure, enhance data sharing and stakeholder collaborations, and encourage the inclusion of a diverse population in clinical research studies.

The Cures Act also mandated the creation of the PRGLAC. This committee included experts from FDA, NIH, National Institute of Mental Health, Veteran's Affairs, medical centers, pharmaceutical companies, universities, CDC, disease-focused associations, and many other governmental, private, for-profit, and nonprofit agencies. The task force's objective was to advise the Secretary of Health and Human Services (HHS) on gaps in knowledge and research on safe and effective therapies for pregnant and lactating women.[8]

In September 2018, 5 months after the release of the FDA guidance, the Task Force submitted its mandated report to the Secretary of HHS that included a detailed analysis of the current environment and specific recommendations that address the issues uncovered during the committee's deliberations. The Secretary is tasked with deciding if, based on the information in the report, changes to regulations or other governmental actions may be warranted to increase knowledge about the findings or to implement changes to policies and practices.[9] This comprehensive report is well worth reading. I have included its executive summary in Appendix II.

The task force's conclusions are similar to those of the FDA's guidance document and the stakeholder opinions that have been covered in this book. The Task Force's 15 priority recommendations are listed in Table 12.1,[10] along with an indicator of its consistency with the FDA guidance and the opinions of industry stakeholders.

TABLE 12.1 Shared recommendations among the Health and Human Services (HHS) Task Force on Research Specific to Pregnant Women and Lactating Women (PRGLAC) Task Force, US Food and Drug Administration (FDA), and industry stakeholders.

HHS PRGLAC recommendations[a]	FDA guidance	Industry stakeholders
1. Include and integrate pregnant women in the clinical research agenda	✓	✓
2. Increase the quantity, quality, and timeliness of research on safety and efficacy of therapeutic products used by pregnant women	✓	✓
3. Expand the workforce of clinicians and research investigators with expertise in obstetric pharmacology and therapeutics	✓	✓
4. Remove regulatory barriers to research in pregnant women	✓	✓

(Continued)

TABLE 12.1 (Continued)

HHS PRGLAC recommendations[a]	FDA guidance	Industry stakeholders
5. Create a public awareness campaign to engage the public and health care providers in research on pregnant women		✓
6. Develop and implement evidence-based communication strategies with health care providers on information relevant to research on pregnant women		✓
7. Develop programs to study therapeutic products used off-patent in pregnant women using the NIHBPCA as a model		
8. Reduce liability to facilitate an evidence base for new therapeutic products that may be used by women who are or may become pregnant		✓
9. Implement a proactive approach to protocol development and study design to include pregnant women in clinical research	✓	✓
10. Develop programs to drive discovery and development of therapeutics and new therapeutic products for conditions specific to pregnant women	✓	✓
11. Utilize and improve existing resources for data to inform the evidence and provide a foundation for research on pregnant women	✓	✓
12. Leverage established and support new infrastructures/collaborations to perform research in pregnant women	✓	✓
13. Optimize registries for pregnancy		✓
14. The Department of Health and Human Services Secretary should consider exercising the authority provided in law to extend the PRGLAC Task Force when its charter expires in March 2019	NA	NA
15. [HHS should] Establish an Advisory Committee to monitor and report on implementation of recommendations, updating regulations, and guidance, as applicable, regarding the inclusion of pregnant women in clinical research		

BPCA, Best Pharmaceuticals for Children Act; NIH, National Institute of Health.
[a]The words "lactating women" were removed from the 15 Task Force recommendations that included them, as neither the FDA guidance nor the stakeholder interviews included this important group of women.

Conclusion

There are many commonalities in the recommendations from pharmaceutical industry stakeholders, the FDA draft guidance document, and the PGRLAC Task Force—including the ethical and medical need to include pregnant women in research, to define the point in the drug development process to enroll them, to employ ethicists and medical specialists for advice and planning, to obtain robust safety and efficacy data from other populations prior to the inclusion of pregnant women, to reduce company liability, etc. But the perspectives of these entities were far apart. While FDA provided some background and ethical context, they focused primarily, as is appropriate to their mandate, on the design and conduct of the studies and the ways in which pregnant women could be protected from research-related risks. The industry's concerns were not so much on how to conduct the studies safely—the researchers have a great deal of expertise in this and felt confident in their ability to design the clinical trials appropriately. Their concerns were more external—legal, financial, and reputational risks. The PRGLAC Task Force Report has taken a broader, oversight view and helps to tie these perspectives together. In my opinion, with the creation of Cure Act's PRGLAC task force, the "political stream" has converged with the "problem stream" and the "policy stream" to create a "policy window of opportunity." The result can be the advancement of social policy if, as Kingdon theorized, this compelling problem, now linked to plausible solutions, will meet the test of political feasibility.[11]

Women's health advocates, medical experts, key informants from industry, and a governmental task force believe that pregnant women and their babies are at a higher risk of adverse medical consequences if they are not included in clinical trials than if they are included in clinical trials.[2,3–5,8] They believe that conducting trials on drug treatment for pregnant women, while ethically, legally, and operationally challenging, is morally required and will be advantageous to pregnant women and their babies, their health care providers and prescribers, and the public health in general. By issuing the draft guidance on the inclusion of pregnant women in clinical research, FDA is challenging industry to confront assumptions and past practices and address the obstacles that prevent effective, evidence-based treatment for pregnant women and their babies.

The broader inclusion of pregnant women in clinical research studies will improve clinical knowledge to decrease inadequate treatment, reduce maternal and neonatal morbidity and mortality, and diminish the unnecessary termination of wanted pregnancies—all consequences of a lack of information about the safety and efficacy of medications used to treat illness during pregnancy.

I support a change in the current practices of the US pharmaceutical industry to broaden the inclusion of pregnant women in clinical studies when appropriate. I found that there is support within the industry for modifications of clinical research inclusion and exclusion criteria. Alternative research designs and other legal, regulatory, and public policy solutions that address sustaining beneficence and reducing litigation risk have been proposed. Improved maternal-infant health outcomes due to the enhanced knowledge of medication efficacy and safety gained from clinical studies in human pregnancies is the goal and would be an important contribution to public health.

The release of the final FDA guidance will provoke industry to respond to the issue, challenge businesses to confront fundamental viewpoints, and spur scientists and researchers to find new ways to contribute to clinical knowledge about the safe and effective treatment of pregnant women who need pharmacotherapy. The business reservations about implementing the necessary changes will need to be addressed. But the wealth of knowledge, passion, and willingness to change that my interviews found within companies to confront this challenge suggest the possibility of change and a potential for improvement in the contribution that the pharmaceutical industry can make to maternal health.

The FDA guidance contains specific recommendations on the planning, approval, implementation, and conduct of clinical trials that enroll pregnant women or who retain women who become pregnant during clinical trials. The book in your hands describes the quantification of pregnant women's actual exclusion from clinical studies that could have ethically included them, and the qualifications (i.e., conditions, reservations, fears) that people in the pharmaceutical industry have about implementing such research. The PRGLAC recommendations will take the issue to the highest levels of government for evaluation and, perhaps, action. From the minutiae of PK testing to the generalities of ethical frameworks, the three documents cover a breadth of material that advocates have been proposing (and industry has been ignoring) while FDA struggled to define and articulate its recommendations. I hope that the publication of this book, the FDA guidance document, and the PRGLAC report will add to the public discourse and will increase the likelihood that the inclusion of pregnant women in clinical trials may proceed ethically, transparently, and with the highest degree of scientific integrity.

Notes

1. Hillebrecht, J. M., McKnight, R. J., Czaban, J. N., & Holian, M. A. (2016). *The writing's on the wall: Wise pharma companies will heed FDA warning in new draft guidance on data integrity cGMPs*. Retrieved from <https://www.dlapiper.com/en/us/insights/publications/2016/05/the-writings-on-the-wall/>.
2. Koren, G., Pastuszak, A., & Ito, S. (1998). Drugs in pregnancy. *New England Journal of Medicine, 338*(16), 1128−1137. https://doi.org/10.1056/NEJM199804163381607.

3. Czeizel, A. E. (1999). The role of pharmacoepidemiology in pharmacovigilance: Rational drug use in pregnancy. *Pharmacoepidemiology and Drug Safety*, *8*(Suppl. 1), S55−S61. https://doi.org/10.1002/(SICI)1099-1557(199904)8:1 + < S55::AID-P > DS404 > 3.0.CO;2-1.

4. Merton, V. (1993). The exclusion of pregnant, pregnable, and once-pregnable people (a.k.a. women) from biomedical research. *American Journal of Law & Medicine*, *19*(4), 369−451 citing Iber, F. L., Riley, W. A., & Murray, P. J. (1987). *Conducting clinical trials*. New York: Plenum Press.

5. Rodger, M. A., Makropoulos, D., Walker, M., Keely, E., Karovitch, A., & Wells, P. S. (2003). Participation of pregnant women in clinical trials: Will they participate and why? *American Journal of Perinatology*, *20*(2), 69−76. https://doi.org/10.1055/s-2003-38318.

6. Wild, V., & Biller-Andorno, N. (2016). Pregnant women's views about participation in clinical research. Chapter 7. In F. Baylis & A. Ballantyne (Eds.), *Clinical research involving pregnant women*. Philadelphia, PA: Springer.

7. *National Institutes of Health. The 20th century cures act.* (2016). Retrieved from <https://www.nih.gov/research-training/medical-research-initiatives/cures>.

8. *U.S. Department of Health and Human Services. Task force on research specific to pregnant women and lactating women, report to secretary, Health and Human Services and Congress.* (2018). Retrieved from <https://www.nichd.nih.gov/sites/default/files/2018-09/PRGLAC_Report.pdf>.

9. *U.S. Department of Health and Human Services. Task force on research specific to pregnant women and lactating women, report to secretary, Health and Human Services and Congress.* (2018). Retrieved from <https://www.nichd.nih.gov/sites/default/files/2018-09/PRGLAC_Report.pdf>.

10. *U.S. Department of Health and Human Services. Task force on research specific to pregnant women and lactating women report to secretary, Health and Human Services and Congress.* (2018). Retrieved from <https://www.nichd.nih.gov/sites/default/files/2018-09/PRGLAC_Report.pdf>.

11. Brieger, W. R. (2006). *Policy making and advocacy.* Retrieved from <http://ocw.jhsph.edu/courses/SocialBehavioralFoundations/PDFs/Lecture19.pdf>.

Appendix I

FDA draft guidance—Pregnant women: scientific and ethical considerations for inclusion in clinical trials

Pregnant Women: Scientific and Ethical Considerations for Inclusion in Clinical Trials
Guidance for Industry

DRAFT GUIDANCE

This guidance document is being distributed for comment purposes only.

Comments and suggestions regarding this draft document should be submitted within 60 days of publication in the *Federal Register* of the notice announcing the availability of the draft guidance. Submit electronic comments to https://www.regulations.gov. Submit written comments to the Dockets Management Staff (HFA-305), Food and Drug Administration, 5630 Fishers Lane, Rm. 1061, Rockville, MD 20852. All comments should be identified with the docket number listed in the notice of availability that publishes in the *Federal Register*.

For questions regarding this draft document, contact the Division of Pediatric and Maternal Health (CDER) at (301) 796-2200 or the Office of Communication, Outreach, and Development (CBER) at 800-835-4709 or 240-402-8010.

U.S. Department of Health and Human Services
Food and Drug Administration
Center for Drug Evaluation and Research (CDER)
Center for Biologics Evaluation and Research (CBER)

April 2018
Clinical/Medical
Revision 1

Pregnant Women: Scientific and Ethical Considerations for Inclusion in Clinical Trials
Guidance for Industry

Additional copies are available from:

Office of Communications, Division of Drug Information
Center for Drug Evaluation and Research
Food and Drug Administration
10001 New Hampshire Ave., Hillandale Bldg., 4th Floor
Silver Spring, MD 20993-0002
Phone: 855-543-3784 or 301-796-3400; Fax: 301-431-6353; Email: druginfo@fda.hhs.gov
https://www.fda.gov/Drugs/GuidanceComplianceRegulatoryInformation/Guidances/default.htm

and/or

Office of Communication, Outreach, and Development
Center for Biologics Evaluation and Research
Food and Drug Administration
10903 New Hampshire Ave., Bldg. 71, rm. 3128
Silver Spring, MD 20993-0002
Phone: 800-835-4709 or 240-402-8010; Email: ocod@fda.hhs.gov
http://www.fda.gov/BiologicsBloodVaccines/GuidanceComplianceRegulatoryInformation/Guidances/default.htm

U.S. Department of Health and Human Services
Food and Drug Administration
Center for Drug Evaluation and Research (CDER)
Center for Biologics Evaluation and Research (CBER)

April 2018
Clinical/Medical
Revision 1

TABLE OF CONTENTS

Contains Nonbinding Recommendations

Draft — Not for Implementation

Pregnant Women: Scientific and Ethical Considerations for Inclusion in Clinical Trials Guidance for Industry[1]

I. INTRODUCTION

This guidance provides recommendations about how and when to include pregnant women in drug development clinical trials for drugs and biological products based on the Food and Drug Administration's (FDA's or Agency's) current thinking on this subject.[2] Specifically, this guidance supports an informed and balanced approach to gathering data on the use of drugs and biological products during pregnancy through judicious inclusion of pregnant women in clinical trials and careful attention to potential fetal risk. This draft guidance is intended to serve as a focus for continued discussions among various entities such as the Agency, pharmaceutical manufacturers, the academic community, institutional review boards (IRBs), and others who are involved with the conduct of clinical trials in pregnant women.[3]

This guidance discusses the scientific and ethical issues that should be addressed when considering the inclusion of pregnant women in drug development clinical trials. From a scientific and ethical standpoint, the population of pregnant women is complex based on the interdependency of maternal and fetal well-being, and the need to take into consideration the risks and benefits of a drug to both woman and fetus (American College of Obstetricians and Gynecologists 2015). The scientific and ethical issues discussed in this guidance apply both to clinical trials that enroll pregnant subjects and to clinical trials that allow enrolled subjects who become pregnant to remain in the trial.

---

[1] This guidance has been prepared by the Division of Pediatric and Maternal Health in the Center for Drug Evaluation and Research (CDER) in cooperation with the Center for Biologics Evaluation and Research and the Office of Good Clinical Practice, Office of Special Medical Programs, in the Office of the Commissioner at the Food and Drug Administration.

[2] Throughout this guidance, the term *drug* means drug and biological products regulated by CDER or CBER.

[3] In addition to consulting guidances, sponsors are encouraged to contact the appropriate review division to discuss specific issues that arise during drug development.

Some of the information provided in this guidance applies to drugs indicated to treat pregnancy-specific conditions (e.g., preterm labor, pre-eclampsia), but the larger focus is on drugs indicated for conditions that occur commonly among females of reproductive potential. Women in this group may require treatment for chronic disease or acute medical problems, and may become pregnant multiple times during the reproductive phase of their lives.

This guidance does not discuss general clinical trial design issues or statistical analysis. Those topics are addressed in the ICH guidances for industry *E9 Statistical Principles for Clinical Trials*, *E10 Choice of Control Group and Related Issues in Clinical Trials*,[4] and the draft ICH guidance for industry *E9(R1) Statistical Principles for Clinical Trials: Addendum: Estimands and Sensitivity Analysis in Clinical Trials*.[5] The draft guidance for industry *Pharmacokinetics in Pregnancy — Study Design, Data Analysis, and Impact on Dosing and Labeling*[6] and certain disease-specific and drug class-specific guidances may provide additional considerations for studying pregnant women during drug development.

In general, FDA's guidance documents do not establish legally enforceable responsibilities. Instead, guidances describe the Agency's current thinking on a topic and should be viewed only as recommendations, unless specific regulatory or statutory requirements are cited. The use of the word *should* in Agency guidances means that something is suggested or recommended, but not required.

II.   BACKGROUND

In the interests of promoting maternal/fetal health and informed prescribing decisions during pregnancy, this guidance addresses the challenges of including pregnant women in drug development research. There are more than 60 million women in the United States between the ages of 15 and 44 years, and almost 4 million births per year (U.S. National Vital Statistics Reports). Like women who are not pregnant, some pregnant women need to use drugs to manage chronic disease conditions or treat acute medical problems. To the extent there is labeling information for pregnant women, it is usually based on nonclinical data with or without limited human safety data. The frequent lack of information based on clinical data often leaves the health care provider (HCP) and the patient reluctant to treat the underlying condition, which in some cases may result in more harm to the woman and the fetus than if she had been treated. In addition, pregnant women often use medically necessary drugs without a clear scientific understanding of the risks and benefits to themselves or their developing fetuses (Lyerly et al. 2008).

---

[4] We update guidances periodically. To make sure you have the most recent version of a guidance, check the FDA Drugs or Biologics guidance web page at
https://www.fda.gov/Drugs/GuidanceComplianceRegulatoryInformation/Guidances/default.htm or
https://www.fda.gov/BiologicsBloodVaccines/GuidanceComplianceRegulatoryInformation/default.htm.

[5] When final, this guidance will represent the FDA's current thinking on this topic.

[6] When final, this guidance will represent the FDA's current thinking on this topic.

Currently, information about drug use in pregnancy generally is collected in the postmarketing setting, using data from observational studies such as pregnancy exposure registries and other cohort studies, case control studies, and surveillance methods. Historically, there have been barriers to obtaining data from pregnant women in clinical trials in an effort to protect them and their fetuses from research-related risks. However, in certain situations, it may be helpful to collect data in pregnant women in the setting of a clinical trial (Goldkind et al. 2010). For example, it may be useful to compare the safety and efficacy of a drug that has been considered the standard of care for pregnant women with a newer treatment (Jones et al. 2010). In other situations, a woman's health and the well-being of her fetus may benefit from clinical trial participation. For example, a pregnant woman may need access to experimental therapies in a clinical trial setting because there are no approved treatment options available. Sometimes a drug treatment offered only through a clinical trial will hold out the prospect of direct benefit to the pregnant woman and/or her fetus beyond otherwise available therapies. For example, some clinical trials for drugs that treat human immunodeficiency virus (HIV), tuberculosis, and malaria enroll pregnant women (or provide that patients who become pregnant can continue enrollment) based on ethical principles and clinical need.

There are multiple reasons for considering the inclusion of pregnant women in clinical trials, including the following:

- Women need safe and effective treatment during pregnancy

- Failure to establish the dose/dosing regimen, safety, and efficacy of treatments during pregnancy may compromise the health of women and their fetuses

- In some settings, enrollment of pregnant women in clinical trials may offer the possibility of direct benefit to the woman and/or fetus that is unavailable outside the research setting

- Development of accessible treatment options for the pregnant population is a significant public health issue

Extensive physiological changes associated with pregnancy may alter drug pharmacokinetics and pharmacodynamics, which directly affects the safety and efficacy of a drug administered to a pregnant woman through alterations in drug absorption, distribution, metabolism, and excretion.[7] Pregnancy-related changes in various organ systems (e.g., gastrointestinal, cardiovascular, and renal) also may alter drug pharmacokinetics and pharmacodynamics. For example, a 30 to 40 percent increase in glomerular filtration rate results in much higher rates of clearance for some drugs during pregnancy (Mattison and Zajicek 2006); therefore, prescribing often occurs in the absence of knowledge regarding the dose required to achieve the desired therapeutic effect (Andrew et al. 2007).

---

[7] See the draft guidance for industry *Pharmacokinetics in Pregnancy — Study Design, Data Analysis, and Impact on Dosing and Labeling.*

Filling the knowledge gaps regarding safe and effective use of drugs in pregnant women is a critical public health need, but one that raises complex issues.

III.   ETHICAL CONSIDERATIONS

The inclusion of pregnant women in clinical trials is guided by human subject protection regulations and involves complex risk-benefit assessments that vary depending on the seriousness of the disease, the availability of other treatments, the trial design, and whether the proposed investigation will occur in the premarketing or postmarketing setting. Because of the complex ethical issues involved in designing clinical trials that include pregnant women, sponsors should consider including an ethicist in planning their drug development programs. Moreover, sponsors should consider meeting with the appropriate FDA review division early in the development phase to discuss when and how to include pregnant women in the drug development plan. These discussions should involve FDA experts in bioethics and maternal health.

A.   FDA Regulations That Govern Research in Pregnant Women

FDA-regulated clinical trials in pregnant women must conform to all applicable FDA regulations, including those related to human subject protections (21 CFR part 56, Institutional Review Boards, and 21 CFR part 50, subpart B, Informed Consent of Human Subjects). In addition, if the trial is supported or conducted by the Department of Health and Human Services (HHS), then 45 CFR part 46 may also apply, which would include subpart B, Additional Protections for Pregnant Women, Human Fetuses and Neonates Involved in Research.[8] The FDA regulations do not contain a section similar to 45 CFR part 46, subpart B; however, the FDA recommends that these requirements be satisfied for FDA-regulated clinical research. Subpart B requires that trials supported or conducted by HHS meet all of the following 10 conditions:

1. Where scientifically appropriate, nonclinical studies, including studies on pregnant animals, and clinical studies, including studies on nonpregnant women, have been conducted and provide data for assessing potential risks to pregnant women and fetuses;

2. The risk to the fetus is caused solely by interventions or procedures that hold out the prospect of direct benefit for the woman or the fetus; or, if there is no such prospect of benefit, the risk to the fetus is not greater than minimal[9] and the purpose of the research is the development of important biomedical knowledge which cannot be obtained by any other means;

3. Any risk is the least possible for achieving the objectives of the research;

---

[8] See 45 CFR 46.204.

[9] See section III.B., Research-Related Risks, for discussion of minimal risk.

Draft — Not for Implementation

4. The pregnant woman's consent is obtained in accord with the informed consent provisions of 45 CFR part 46, subpart A;

5. If the research holds out the prospect of direct benefit solely to the fetus then the consent of the pregnant woman and the father is obtained in accord with the informed consent provisions of 45 CFR part 46, subpart A, except that the father's consent need not be obtained if he is unable to consent because of unavailability, incompetence, or temporary incapacity or the pregnancy resulted from rape or incest;

6. Each individual providing consent is fully informed regarding the reasonably foreseeable impact of the research on the fetus or neonate;

7. For children as defined in § 46.402(a) who are pregnant, assent and permission are obtained in accord with the provisions of 45 CFR part 46, subpart D;

8. No inducements, monetary or otherwise, will be offered to terminate a pregnancy;

9. Individuals engaged in the research will have no part in any decisions as to the timing, method, or procedures used to terminate a pregnancy; and

10. Individuals engaged in the research will have no part in determining the viability of a neonate.

IRBs are required to possess the professional competence necessary to review the specific research activities that they oversee (21 CFR 56.107(a)). IRBs must include persons who are knowledgeable in areas about the acceptability of proposed research in terms of institutional commitments and regulations, applicable law, and standards of professional conduct and practice (21 CFR 56.107(a)). Therefore, if an IRB regularly reviews research involving pregnant women, the IRB must consider including one or more individuals who are knowledgeable about and experienced in working with such subjects (21 CFR 56.107(a)). When an IRB considers whether to approve a protocol involving pregnant women, it should consider only those risks and benefits (direct to the subjects, or generalizable knowledge) that may result from the research itself (as distinguished from risks and benefits of therapies that subjects would receive even if not participating in the research) (21 CFR 56.111(a)(2)). Additionally, IRBs are required to determine that additional safeguards are included in the trial to protect the rights and welfare of subjects who are pregnant (21 CFR 56.111(b)).

Additional issues are raised by pregnant minors. Depending on state law, a pregnant minor may be considered emancipated by virtue of her pregnancy, a mature minor, or still a child (see the definition of children under 21 CFR 50.3(o)). IRBs should be familiar with applicable law of the jurisdiction in which a trial will be conducted. In the event that a clinical trial regulated by the FDA allows the enrollment of pregnant minors, or a minor becomes pregnant while enrolled in a clinical trial, and the pregnant minor meets the definition of a child under applicable state law, the IRB would have to comply with the applicable requirements of 21 CFR part 50, subpart D, Additional Safeguards for Children in Clinical Investigations.

B.   Research-Related Risks

Research-related risks may meet the regulatory definition for *minimal risk* or may involve greater than minimal risk. FDA regulations define minimal risk as follows (21 CFR 50.3(k)):

> "*Minimal risk* means that the probability and magnitude of harm or discomfort anticipated in the research are not greater in and of themselves than those ordinarily encountered in daily life or during the performance of routine physical or psychological examinations or tests."

Research-related risks are the risks specifically associated with the trial interventions or procedures. If a woman is assigned to receive a drug while enrolled in a clinical trial (i.e., the assignment of the drug is determined by the protocol), then the risks associated with the drug would be considered research-related.

In contrast, risks are not research-related when they are independent of the study and not associated with a trial intervention or protocol requirements. In other words, when a study collects data about drug treatment during pregnancy but the drug was prescribed before study enrollment by the patient's HCP, then the risks associated with the drug use are not research-related risks (Sheffield et al. 2014). For example, in a study in which the investigator plans to assess the pharmacokinetics of a particular selective serotonin reuptake inhibitor (SSRI) during pregnancy, the investigator enrolls pregnant women with a history of major depression who are currently managed on this drug. In this study the SSRI does not create research-related risk, because the patients are already using the SSRI (as previously prescribed by their HCPs) to manage their medical conditions. The only risks of the study are those associated with study-specific procedures (e.g., blood sample collection), and potential loss of confidentiality or privacy.

In this situation, the research-related risk to the fetus is minimal, and the purpose of the research is the development of important biomedical knowledge, which cannot be obtained by any other means. Some dedicated pharmacokinetic (PK) studies conducted with pregnant women (such as the previous SSRI example) can offer direct benefit to subjects if the data are used during the trial to adjust the dosing for individual subjects when clinically appropriate. The informed consent process should include discussion of expectations about whether trial data will be monitored and evaluated in a way that can potentially benefit the subject during the trial.

There may be circumstances in which a clinical trial can potentially expose a fetus to greater than minimal risk. Pregnant women can be enrolled in clinical trials that involve greater than minimal risk to the fetuses if the trials offer the potential for direct clinical benefit to the enrolled pregnant women and/or their fetuses. For example, this benefit may result from access to: (1) a needed but otherwise unavailable therapy (e.g., a new antituberculosis drug for multidrug resistant disease); or (2) a drug or biologic that reduces the risk for acquiring a serious health condition (e.g., a vaginal microbicide that reduces transmission of HIV and herpes simplex virus).

C.   General Guidelines for Including Pregnant Women in Clinical Trials

This section provides general guidelines and considerations for including pregnant women in clinical trials. However, every drug development situation is unique, and individualized approaches to clinical trial design may be required to facilitate inclusion of pregnant women in specific drug development plans.

The FDA considers it ethically justifiable to include pregnant women with a disease or medical condition requiring treatment in clinical trials under the following circumstances:

In the postmarketing setting (i.e., FDA-approved drugs)

- Adequate nonclinical studies (including studies on pregnant animals) have been completed[10]

  and

- There is an established safety database in nonpregnant women from clinical trials or preliminary safety data from the medical literature and/or other sources regarding use in pregnant women

  and one of the following:

- Efficacy cannot be extrapolated

  and/or

- Safety cannot be assessed by other study methods

In the premarketing setting (i.e., investigational drugs)

- Adequate nonclinical studies (including studies on pregnant animals) have been completed

  and

- The clinical trial holds out the prospect of direct benefit to the pregnant woman and/or fetus that is not otherwise available outside the research setting or cannot be obtained by any other means (e.g., the pregnant woman may not have responded to other approved treatments or there may not be any treatment options)

The above conditions would also apply to a drug that is being developed to treat a pregnancy-specific condition.

---

[10] The phrase *adequate nonclinical studies* refers to recommendations for the design and conduct of reproductive toxicology and other nonclinical studies described in the ICH guidances for industry *M3(R2) Nonclinical Safety Studies for the Conduct of Human Clinical Trials and Marketing Authorization for Pharmaceuticals* and *S5(R2) Detection of Toxicity to Reproduction for Medicinal Products: Addendum on Toxicity to Male Fertility.*

Women who become pregnant while enrolled in a clinical trial

When a pregnancy has been identified during a clinical trial, unblinding should occur so that counseling may be offered based on whether the fetus has been exposed to the investigational drug, placebo, or control. The risks and benefits of continuing versus stopping investigational treatment can be reviewed with the pregnant woman. Pregnant women who choose to continue in the clinical trial should undergo a second informed consent process that reflects these additional risk-benefit considerations.

If fetal exposure has already occurred, a woman who becomes pregnant while enrolled in a clinical trial should be allowed to continue on the investigational drug if the potential benefits of continued treatment for the woman outweigh the risks of ongoing fetal exposure to the investigational drug, of discontinuing maternal therapy, and/or of exposing the fetus to additional drugs if placed on an alternative therapy. Regardless of whether the woman continues in the trial, it is important to collect and report the pregnancy outcome.

IV.   OTHER CONSIDERATIONS

Including pregnant women in a trial involves careful risk-benefit assessments. All trials must be designed to minimize risk as much as possible while preserving the ability to achieve the objectives of the research (21 CFR 56.111). Some general considerations for sponsors and investigators include:

- Obtaining adequate reproductive and developmental toxicology data in relevant nonclinical models

- Identifying the trial population that will derive the most benefit while trying to minimize risk

- Considering the gestational timing of exposure to the investigational drug in relation to fetal development

- Choosing appropriate control populations

Sponsors should also consider the issues discussed in the following sections when designing a clinical trial that will include pregnant women.

A.   Disease Type and Availability of Therapeutic Options in the Pregnant Population

Sponsors should take into account the incidence of the disease, the severity of the disease (e.g., whether or not it is life-threatening), and the availability of other therapeutic options and their risks. Pregnant patients with no other viable therapeutic options (e.g., drug resistance, drug intolerance, contraindication, drug allergy) to treat a serious or life-threatening disease or condition may be appropriate candidates to enroll in a clinical trial.

B.   Timing of Enrollment

The most appropriate time to include pregnant women in clinical trials during drug development may differ. Nonclinical reproductive and developmental toxicology studies generally should be completed before enrolling pregnant women in clinical trials.[11] In general, phase 1 and phase 2 clinical trials in a nonpregnant population that include females of reproductive potential should be completed before sponsors enroll pregnant women in later phase clinical trials. Sponsors should consider whether any of the following situations apply in determining when to enroll pregnant women in the drug development process.

- *If there are limited safety data or other approved (i.e., safe and effective) treatments are available*: In this situation, it may be more appropriate to complete phase 3 clinical trials in a nonpregnant population before enrolling pregnant women and exposing them to the investigational drug

- *If there are limited therapeutic options*: In these situations, the risk-benefit considerations may favor enrollment of pregnant women in earlier phase trials

- *If there are safety data for a drug that has been studied previously for other indications or populations*: In these situations, the risk-benefit considerations may favor enrollment of pregnant women in earlier phase trials

C.   Pharmacokinetic Data

Because of the extensive physiological changes associated with pregnancy, PK parameters may change, sometimes enough to justify changes in dose or dosing regimen. For drug development programs where there are plans to enroll pregnant women in a phase 3 clinical trial, PK data in pregnant women should be collected during the phase 2 clinical trials to guide appropriate dosing in phase 3. In situations where pregnant women are enrolled in phase 3 clinical trials for a marketed drug, PK data should be collected as part of the trial.

In appropriate situations, nonpregnant women who become pregnant while on the investigational drug and consent to remain on the drug can also consent to PK assessments at steady state to collect data on correct dosing during pregnancy. Modeling and simulation have been increasingly used to support the design of clinical PK studies (Xia et al. 2013; Ke et al. 2013). For PK studies including pregnant patients, physiological changes during and after pregnancy that are critical for drug absorption and disposition may need to be considered in the model.

For additional information on PK modeling, study design considerations, and PK studies in pregnant women, refer to the draft guidance for industry *Pharmacokinetics in Pregnancy — Study Design, Data Analysis, and Impact on Dosing and Labeling.*

---

[11] See ICH M3(R2).

D.   Safety Data Collection and Monitoring

When pregnant women are enrolled in a clinical trial, data collection elements should include, at a minimum: gestational age at enrollment; gestational timing and duration of drug exposure; and pregnancy outcomes including adverse maternal, fetal, and neonatal events. Enrolled pregnant patients should also receive obstetrical care that meets the recognized standards of care. Infants born to mothers who were exposed to the investigational drug should have follow-up safety information collected. Systemic drug exposure to the fetus/newborn can be evaluated by collecting cord blood or neonatal levels of drug and/or metabolites, depending on the timing of exposure to the drug and its half-life.

Clinical trials that enroll pregnant women should include investigators or consultants who have expertise in obstetrics and/or maternal/fetal medicine, depending on the underlying conditions treated by the investigational drug.

All clinical trials require monitoring (21 CFR 312.50 and 312.56), and no single approach to monitoring is appropriate or necessary for every clinical trial.[12] Clinical trials that involve pregnant women should include a data monitoring plan that includes members with relevant specialty and perinatal expertise to permit ongoing recognition and evaluation of safety concerns that arise during the course of the trial. This facilitates appropriate, expert assessment of adverse event reports.

E.   Stopping a Clinical Trial That Enrolls Pregnant Women

There may be situations where it would be appropriate to stop a randomized, controlled clinical trial that is enrolling pregnant women. Examples include the following:

- An appropriately planned interim analysis demonstrates superior efficacy of the control or active comparator arm.

- There are documented serious maternal or fetal adverse events that can be reasonably attributed to drug exposure and are deemed to exceed the potential benefits of drug treatment. This determination should include consideration of alternative effective treatments and the risks of the underlying condition.

---

[12] See the guidance for clinical trial sponsors *Establishment and Operation of Clinical Trial Data Monitoring Committees* and the guidance for industry *Oversight of Clinical Investigations — A Risk-Based Approach to Monitoring.*

REFERENCES

American College of Obstetricians and Gynecologists, 2015, Ethical Considerations for Including Women as Research Participants, Committee Opinion No. 646, November.

Andrew, MA, TR Easterling, DB Carr, D Shen, ML Buchanan, T Rutherford, R Bennett, P Vicini, and MF Hebert, 2007, Amoxicillin Pharmacokinetics in Pregnant Women: Modeling and Simulations of Dosage Strategies, Clin Pharm and Therapeutics, 81(4):547–556.

Goldkind, S, L Sahin, and B Gallauresi, 2010, Enrolling Pregnant Women in Research — Lessons From the H1N1 Influenza Pandemic, N Engl J Med, Jun 17, 362(24):2241–2243.

Jones, H, K Kaltenbach, S Heil, S Stine, M Coyle, A Arria, K O'Grady, P Selby, P Martin, and G Fischer, 2010, Neonatal Abstinence Syndrome After Methadone or Buprenorphine Exposure, N Engl J Med, Dec 9, 363(24):2320–2331.

Ke, AB, SC Nallani, P Zhao, A Rostami-Hodjegan, and J Unadkat, 2013, Expansion of a PBPK Model to Predict Disposition in Pregnant Women of Drugs Cleared Via Multiple CYP Enzymes, Including CYP2B6, CYP2C9, and CYP2C19, BJCP, 77:3:554–570.

Lyerly, AD, MO Little, and R Faden, 2008, The Second Wave: Toward Responsible Inclusion of Pregnant Women in Research, Int J Feminist Approaches to Bioethics, Fall, 1(2):5–22.

Mattison, D and A Zajicek, 2006, Gaps in Knowledge in Treating Pregnant Women, Gend Med, Sep, 3(3):169–82.

Sheffield, JS, D Siegel, M Mirochnick, RP Heine, C Nguyen, K Bergman, RM Savic, J Long, KE Dooley, and M Nesin, 2014, Designing Drug Trials: Considerations for Pregnant Women, Clinical Infectious Diseases, 59(S7):S437–S444.

U.S. National Vital Statistics Reports Rapid Release, 2017, Births: Provisional Data for 2016, Report No. 002, June.

Xia, B, T Heimbach, R Gollen, C Nanavati, and H He, 2013, A Simplified PBPK Modeling Approach for Prediction of Pharmacokinetics of Four Primarily Renally Excreted and CYP3A Metabolized Compounds During Pregnancy, AAPSJ, 15(4):1012–1024.

Appendix II

Executive summary: task force on research specific to pregnant women and lactating women

Report to

Secretary, Health and Human Services

Congress

September 2018

Read the full report at: https://www.nichd.nih.gov/sites/default/files/2018-09/PRGLAC_Report.pdf

Executive summary

The Task Force on Research Specific to Pregnant Women and Lactating Women ("Task Force" or "PRGLAC") was established by section 2041 of the 21st Century Cures Act, P.L. 114-255 (report, Section 6: Appendix I) and convened in accordance with the Federal Advisory Committee Act (5 U. S.C. App.) with membership as outlined in the Act (report, Section 6: Appendix II). The Task Force was charged with providing advice and guidance to the Secretary of Health and Human Services (HHS) on activities related to identifying and addressing gaps in knowledge and research on safe and effective therapies for pregnant women and lactating women, including the development of such therapies and the collaboration on and coordination of such activities. In addition to advising the Secretary, the Task Force was charged with preparing and submitting to the Secretary and Congress a report that includes five elements (Box 1).

The Task Force developed 15 recommendations (report, Section 5) based on information gleaned during four open meetings and a public comment

BOX 1 Statutory charge to the Task Force on Research Specific to Pregnant Women and Lactating Women, as established by the 21st Century Cures Act.

…Provide advice and guidance to the Secretary of the Department of Health and Human Services and prepare a report, for the Secretary to transmit to Congress, that includes the following:

1. A plan to identify and address gaps in knowledge and research regarding safe and effective therapies for pregnant women and lactating women, including the development of such therapies.
2. Ethical issues surrounding the inclusion of pregnant women and lactating women in clinical research.
3. Effective communication strategies with health care providers and the public on information relevant to pregnant women and lactating women.
4. Identification of Federal activities, including:
   a. the state of research on pregnancy and lactation;
   b. recommendations for the coordination of, and collaboration on research related to pregnant women and lactating women;
   c. dissemination of research findings and information relevant to pregnant women and lactating women to providers and the public; and
   d. existing Federal efforts and programs to improve the scientific under-standing of the health impacts on pregnant women, lactating women, and related birth and pediatric outcomes, including with respect to phar-macokinetics, pharmacodynamics, and toxicities.
5. Recommendations to improve the development of safe and effective thera-pies for pregnant women and lactating women.

period (report, Section 6: Appendices III–V and IX). This Executive Summary provides context for the recommendations (Box 2). A central theme resonated throughout the recommendations—the need to alter cultural assumptions that have significantly limited scientific knowledge of therapeu-tic product safety, effectiveness, and dosing for pregnant and lactating women. The cultural shift is necessary to emphasize the importance and pub-lic health significance of building a knowledge base to inform medical decision-making for these populations.

Consequently, research on therapies for these populations must be facili-tated and greatly augmented.

Over six million women are pregnant in the United States each year. Of these women, more than 90% take at least one medication during pregnancy and lactation (report, Section 6: Appendix VI). However, pregnant women and lactating women are often excluded from clinical research that could ultimately help these populations. The Task Force recommends that this tra-jectory of exclusion be altered to include and integrate pregnant women and

BOX 2 Task Force on Research Specific to Pregnant Women and Lactating Women Recommendations.

1. Include and integrate pregnant women and lactating women in the clinical research agenda.
2. Increase the quantity, quality, and timeliness of research on safety and efficacy of therapeutic products used by pregnant women and lactating women.
3. Expand the workforce of clinicians and research investigators with expertise in obstetric and lactation pharmacology and therapeutics.
4. Remove regulatory barriers to research in pregnant women.
5. Create a public awareness campaign to engage the public and health care providers in research on pregnant women and lactating women.
6. Develop and implement evidence-based communication strategies with health care providers on information relevant to research on pregnant women and lactating women.
7. Develop separate programs to study therapeutic products used off-patent in pregnant women and lactating women using the National Institute of Health (NIH) Best Pharmaceuticals for Children Act (BPCA) as a model.
8. Reduce liability to facilitate an evidence base for new therapeutic products that may be used both by women who are or may become pregnant and by lactating women.
9. Implement a proactive approach to protocol development and study design to include pregnant women and lactating women in clinical research.
10. Develop programs to drive discovery and development of therapeutics and new therapeutic products for conditions specific to pregnant women and lactating women.
11. Utilize and improve existing resources for data to inform the evidence and provide a foundation for research on pregnant women and lactating women.
12. Leverage established—and support new—infrastructures/collaborations to perform research in pregnant women and lactating women.
13. Optimize registries for pregnancy and lactation.
14. The Department of Health and Human Services Secretary should consider exercising the authority provided in law to extend the PRGLAC Task Force when its charter expires in March 2019.
15. Establish an Advisory Committee to monitor and report on implementation of recommendations, updating regulations, and guidance, as applicable, regarding the inclusion of pregnant women and lactating women in clinical research.

lactating women in the clinical research agenda (*Recommendation 1*). To date, their exclusion may be motivated by concern about possible harms of medication use during pregnancy or lactation. Although the potential harms of unmedicated disease for both the woman and the developing fetus or breast-fed newborn usually elicit less study, they are nonetheless important. A comprehensive review of research in recent years conducted for the Task Force deliberations clearly showed the extremely limited information available on medication use in pregnancy and lactation (report, Section 6: Appendix VI).

Anecdotal reports state that many pregnant women and lactating women also use herbal and dietary supplements, but there are limited to no data to inform their use, dosing, or therapeutic levels. Since these dietary supplements are regulated differently than drugs by the United States Food and Drug Administration (FDA) and do not require premarket approval, our ability to understand the safety and efficacy of what women consume during pregnancy and lactation is limited.

Evidence-based answers are required for women and their clinicians to make fully informed choices based on the risks and benefits of medicating or not medicating conditions during pregnancy and lactation. The provision of clinical data is essential to increasing the quantity, quality, and timeliness of research on safety and efficacy of therapeutic products used by pregnant women and lactating women (*Recommendation 2*).

Furthermore, expansion of this research requires the training and career development of a workforce with expertise in obstetric and lactation pharmacology and therapeutics (*Recommendation 3*).

The unique relationship between a pregnant woman and her fetus, or a lactating woman and her child, has resulted in federal regulations, guidance, and rules aimed at protecting the woman, fetus, and/or child (report, Section 2). A universal consent for pregnancy and lactation studies might be preferable given state law differences (including those affecting minors). Subpart B of the Common Rule regulations (Section 46.204(e)) currently requires both maternal and paternal consent for a pregnant woman to take part in a study that benefits only the fetus. Changing this to a requirement for maternal consent only would facilitate participation in research, more consistent with the requirements for pediatrics, under Subpart D 50.51 and 50.52, where single parent consent is acceptable. Lactating women may not face the same hurdles as pregnant women in joining a research study, since it may be easier to predict risk in this population.

The Task Force recommends removal of unnecessarily burdensome regulatory barriers to research involving pregnant women (*Recommendation 4*).

For any cultural shift to occur, behavior change requires targeted communication strategies. When communicating information relevant to treating disorders in pregnant women and lactating women, messages must be concise, consistent, tailored, and actionable for health care providers and the public (report, Section 3). The Task Force recommends a focused public awareness campaign that highlights the importance of research on therapeutic products in pregnant women and lactating women, the impact of not taking needed medications during pregnancy and lactation, and the effects of not breastfeeding on the mother and child (*Recommendation 5*). In addition, evidence-based communication strategies with health care providers are needed to increase their knowledge and engagement (*Recommendation 6*).

The Task Force recognized that different approaches for subsets of therapeutic products (drugs, vaccines, and dietary supplements) may be required to make significant progress. These include: (1) therapeutics that are already in use and off-patent, (2) those already in use and on-patent, and (3) those in development for conditions not specific to pregnancy or lactation (e.g., anti-hypertensive agents, antibiotics, and vaccines) and for conditions specific to pregnancy (e.g., hyperemesis) or lactation (e.g., low milk supply). Also, different strategies may be needed to obtain regulatory approval and labeling for therapeutic products for use in pregnant women and in lactating women, with separate processes for prioritization.

The Task Force outlined the parameters for moving forward (report, Section 1, Figure 1).

A major impediment to obtaining data on therapeutic products that are used by pregnant women and lactating women is the concern about liability (report, Section 1, Box 3). Without processes to mitigate liability, it is unlikely that incentives or regulations will provide traction to facilitate the development of an evidence base for therapeutic products that may be used by lactating women or by women who are or may become pregnant. The Task Force recommends implementing a liability-mitigation strategy (*Recommendation 7*). One example is the Vaccine Injury Compensation Program, but for this scenario, the mitigation would need to cover liability regardless of whether the therapeutic product receives marketing approval or not. In addition, targeted programs and/or strengthening the FDA's authority to require, as part of applications for approval of new therapeutics, clinically relevant data for pregnant women and lactating women in study designs could expand the evidence base.

Many therapeutic products are already in use by pregnant women and lactating women. These products are commonly off-patent, thus there is no incentive to pharmaceutical manufacturers to obtain safety and efficacy data in pregnant and lactating populations. The National Institutes of Health's (NIH) Best Pharmaceuticals in Children Act program is a successful model for studying off-patent use of therapies in children (report, Section 6: Appendix XI). This model provides specific funding and has a prioritization process to study off-patent therapies that have public health benefit in children but have not been adequately studied. Using elements of this model, the Task Force recommends developing separate programs, one for pregnant women and one for lactating women, with specific funding and prioritization processes to obtain critically needed data (*Recommendation 8*). The Task Force felt it is essential to have separate programs to ensure the appropriate focus on these distinct patient populations and to mitigate concerns of competition between the needs for pregnant women and lactating women if a single program were developed.

Several conditions are specific to pregnancy; however, the pipeline for products to treat these conditions is minimal at best. The first impediment is liability, requiring the aforementioned liability-mitigation strategy (*Recommendation 7*). The second impediment is limited interest in the development of these products despite significant unmet needs. In lactating women, for example, therapeutic products for low milk supply are virtually nonexistent. Pregnant women urgently need therapeutic products for the treatment of preterm labor, hyperemesis, and cholestasis. Because these needs are distinct, separate prioritization processes for products for pregnant women and lactating women are essential. Research programs should be developed to drive discovery and development of new therapeutic products for conditions specific to pregnant women and lactating women (*Recommendation 9*). These programs could include a variety of models, including incentives, requirements, or other methods to stimulate discovery and development.

Examples of targeted programs include the Biomedical Advanced Research and Development Authority, which aims to secure the United States from pandemic influenza and other emerging infectious diseases by moving medical countermeasures, such as vaccines, drugs, and diagnostics, from research through advanced development and consideration for FDA approval. Another example is the NIH vaccine development program, which advances experimental vaccines up to phase II clinical trials.

To support Recommendation 1, the inclusion and integration of pregnant women and lactating women in the clinical research agenda, a proactive approach is needed for protocol development and study design (*Recommendation 10*). Specifically, investigators and sponsors should be required to justify the exclusion of pregnant women and lactating women in their study designs and to develop studies to capture the physiologic changes that occur over time in these populations. To achieve these goals, guidance needs to be developed for both investigators and institutional review boards. Also, the Task Force recommends leveraging and supporting both new and established research networks (report, Section 4 and Section 6: Appendix VII) and collaborations through financial support and incentives to perform this work (*Recommendation 11*). The Task Force also recommends strengthening existing data resources to inform the evidence base and provide a foundation for research on pregnant women and lactating women (*Recommendation 12*). This includes designing health record systems that link mother and infant records, leveraging large databases and data systems, and utilizing innovative methods for analytics.

To date, one of the major methods for obtaining information on pregnant women has been through the use of registries (report, Section 6: Appendix VII). Registries have not been utilized for obtaining data in lactating women.

Registries are typically operated by the pharmaceutical industry, often at the request of the FDA. The Task Force recommends optimizing registries for pregnancy and lactation to include the creation of a user-friendly website for registry listing, developing registry standards with common data elements, and facilitating transparency and access to the data (*Recommendation 13*). Rather than product-specific registries, the Task Force recommends developing disease- or condition-focused registries. The ideal would be a single registry for all therapeutic products. However, establishment of such registries for all relevant conditions will require substantial coordination, collaboration, and funding mechanisms.

Per the 21st Century Cures Act, the charter of the Task Force will expire in March 2019 (Appendix I). Given the large amount of work that remains in order to fully research therapeutic products used by pregnant women and lactating women, the Task Force recommends that the HHS Secretary consider exercising the authority provided in law to extend the Task Force when its charter expires (*Recommendation 14*). Based on the decisions of the Secretary, a continuation of the Task Force may provide more detail on the implementation of the recommendations made to date and address other pertinent areas related to these initial recommendations. Also, the Task Force recommends that the Secretary establish a Federal Advisory Committee to monitor and report on the implementation of recommendations and on updates to regulations and guidance, as applicable, regarding the inclusion of pregnant women and lactating women in clinical research (*Recommendation 15*).

The work of the Task Force augments and extends prior efforts (Appendix X) that recommended the inclusion of pregnant women and lactating women in research. Without research and the establishment of an evidence base, practitioners care for pregnant women and lactating women without adequate data on the safety, efficacy, or appropriate dosing of therapeutic products. Pregnant women and lactating women and their health care providers are left with undesirable options—either taking a therapeutic product without high-quality dosing or safety information or not treating a condition adequately. In the case of lactation, women may be choosing to discontinue breastfeeding to take the therapy based on limited information, which then deprives the mother and infant of the benefits of lactation. Pregnant women, lactating women, their offspring, and families deserve to have this essential information. The Task Force urges the Secretary to take action on these recommendations.

Bibliography

American College of Obstetricians and Gynecologists. (2005). ACOG committee opinion no. 321: Maternal decision making, ethics, and the law. *Obstetrics and Gynecology, 106*(5), 1127–1137.

American College of Obstetricians and Gynecologists. (2007). ACOG committee opinion no. 377: Research involving women. *Obstetrics & Gynecology, 110*(3), 731–736. Available from https://doi.org/10.1097/01.AOG.0000263926.75016.db.

American Public Health Association. (2005). *APHA legislative advocacy handbook: A guide for effective public health advocacy.* Washington, DC: American Public Health Association.

Andrade, S. E., Gurwitz, J. H., & Davis, R. L. (2004). Prescription drug use in pregnancy. *American Journal of Obstetrics and Gynecology, 191*, 398–407. Available from https://doi.org/10.1016/j.ajog.2004.04.025.

Andrew, M. A., Easterling, T. R., Carr, D. B., Shen, D., Buchanan, M., Rutherford, T., ... Hebert, M. F. (2007). Amoxicillin pharmacokinetics in pregnant women: Modeling and simulations of dosage strategies. *Clinical Pharmacology & Therapeutics, 81*(4), 547–556. Available from https://doi.org/10.1038/sj.clpt.6100136.

Appelbaum, P. S., Roth, L. H., Lidz, C. W., Benson, P., & Winslade, W. (1987). False hopes and best data: Consent to research and the therapeutic misconception. *Hastings Center Report, 17*(2), 20–24.

Association of Faculties of Medicine of Canada. (n.d.). Primer on public health population: The policy cycle. In Howlett, M., & Ramesh, M. (1995). *Studying public policy: Policy cycles and policy subsystems.* Toronto: Oxford University Press. Retrieved from <https://phprimer.afmc.ca/en/part-iii/chapter-14/>.

Baylis, F. (2010). Opinion: Pregnant women deserve better. *Nature, 465*, 689–690.

Beauchamp, D. E. (1976). Public health as social justice. *Inquiry, 13*, 101–109.

Beauchamp, T., & Childress, J. (2001). *Principles of biomedical ethics* (5th ed.). New York: Oxford University Press. (1st ed. published in 1979).

Beh, H. (2005). Compensation for research injuries. *IRB: Ethics & Human Research, 27*(3), 11–15. Available from https://doi.org/10.2307/3564074.

Beigi, R. H., Han, K., Venkataramanan, R., Hankins, G. D., Clark, S., Herbert, M. F., ... Caritis, S. N. (2011). Pharmacokinetics of ostemavir among pregnant and nonpregnant women. *American Journal of Obstetrics & Gynecology, 204*(6 Suppl. 1), S84–S88. Available from https://doi.org/10.1016/j.ajog.2011.03.002.

Beran, R. G. (2006). The ethics of excluding women who become pregnant while participating in clinical trials of anti-epileptic medications. *Seizure, 15*, 563–570. Available from https://doi.org/10.1016/j.seizure.2006.08.008.

Berlin, J. A., & Ellenberg, S. S. (2009). Commentary: Inclusion of women in clinical trials. *BMC Medicine, 7*(56), 1–3. Available from https://doi.org/10.1186/1741-7015-7-56.

Brent, R. L. (1995). Bendectin: Review of the medical literature of a comprehensively studied human nonteratogen and the most prevalent tortogen-litigen. *Reproductive Toxicology Review, 9*(4), 337–349. Available from https://doi.org/10.1016/0890-6238(95)00020-B.

Brent, R. L. (2004a). Environmental causes of human congenital malformations: The pediatrician's role in dealing with these complex clinical problems caused by a multiplicity of environmental and genetic factors. *Pediatrics, 113*(4), 957–968.

Brent, R. L. (2004b). Utilization of animal studies to determine the effects and human risks of environmental toxicants. *Pediatrics, 113*, 984–995.

Brent, R. L. (2007). How does a physician avoid prescribing drugs and medical procedures that have reproductive and developmental risks? *Clinics in Perinatology, 34*, 233–262.

Brieger, W. R. (2006). *Policy making and advocacy.* Johns Hopkins University. Retrieved from <http://ocw.jhsph.edu/courses/SocialBehavioralFoundations/PDFs/Lecture19.pdf>.

Brody, J. (1983, June 19). Shadow of doubt wipes out Bendectin. *New York Times.*

Burton, B. K., & Dunn, C. P. (1996). Feminist ethics as moral grounding for stakeholder theory. *Business Ethics Quarterly, 6*(2), 133–147.

Cain, J., Lowell, J., Thorndyke, L., & Localio, A. R. (2000). Contraceptive requirements for clinical research. *Obstetrics & Gynecology, 95*(6), 861–866. Available from https://doi.org/10.2337/dc15-2723.

Carter, L. (2002). *A primer to ethical analysis, Office of Public Policy and Ethics, University of Queensland; adapted from Beauchamp & Childress. Principles of biomedical ethics* (5th ed.). Oxford University Press.

Center for Disease Control and Prevention. *National vital statistics report: International comparisons of infant mortality and related factors: United States and Europe, 2010.* (2014). Retrieved from <https://www.cdc.gov/nchs/data/nvsr/nvsr63/nvsr63_05.pdf>.

Centers for Disease Control and Prevention. *National Center for Health Statistics: Infant health.* (2017). Retrieved from <https://www.cdc.gov/nchs/fastats/infant-health.htm>.

Centers for Disease Control and Prevention. *Reproductive health: Pregnancy mortality surveillance system.* (2018a). Retrieved from <https://www.cdc.gov/reproductivehealth/maternalinfanthealth/pregnancy-mortality-surveillance-system.htm>.

Centers for Disease Control and Prevention. (2018b). *Birth defects.* Washington, DC: Author. Retrieved from http://www.cdc.gov/ncbddd/birthdefects/data.html.

Charo, R. A. (1993). Protecting us to death: Women, pregnancy, and clinical research trials. *Saint Louis University Law Journal, 38*, 135–187.

Chopra, S. S. (2003). Industry funding of clinical trials: Benefit or bias? *Journal of the American Medical Association, 290*(1), 113–114. Available from https://doi:10.1001/jama.290.1.113.

Clayton, E. W. (1994). Liability exposure when offspring are injured because of their parents' participation in clinical trials. In C. C. Mastroianni, R. R. Faden, & D. D. Federman (Eds.), *Women and health research: Workshop and commissioned papers.* Washington, DC: National Academies Press.

Comment From Multiple Signatories Anonymous. *Comment on the FDA notice: Pregnant women: Scientific and ethical considerations for inclusion in clinical trials; draft guidance. ID: FDA-2018-D-1201-0038.* (2018). Retrieved from <https://www.regulations.gov/document?D = FDA-2018-D-1201-0038>.

Conceptual Framework. (2009). *Mosby's medical dictionary* (8th ed.). Retrieved from <https://medical-dictionary.thefreedictionary.com/conceptual + framework>.

Constantine, M. M. (2014). Physiologic and pharmacokinetic changes in pregnancy. *Frontiers in Pharmacology, 5*, 65. Available from https://doi.org/10.3389/fphar.2014.00065.

Cooper, W. O., Hernandez-Diaz, S., Arbogast, P. G., Dudley, J. A., Dyer, S., Gideon., et al. (2006). Major congenital malformations after first-trimester exposure to ACE inhibitors. *New England Journal of Medicine, 354*, 2443−2451. Available from https://doi.org/ 10.1056/NEJMoa055202.

Council for International Organizations of Medical Sciences. *CIOMS international ethical guidelines for health-related research involving humans.* (2012). Retrieved from <https://cioms.ch/ wpcontent/uploads/2016/08/International_Ethical_Guidelines_for_Biomedical_Research_ Involving_Human_Subjects.pdf>.

Council for International Organizations of Medical Sciences. *CIOMS international ethical guidelines for health-related research involving humans.* (2016). Retrieved from <https://cioms. ch/wp-content/uploads/2017/01/WEB-CIOMS-EthicalGuidelines.pdf>.

Czeizel, A. E. (1999). The role of pharmacoepidemiology in pharmacovigilance: Rational drug use in pregnancy. *Pharmacoepidemiology and Drug Safety, 8*(Supp. 1), S55−S61. Available from https://doi.org/10.1002/(SICI)1099-1557(199904)8:1 + < S55::AID-P > DS404 > 3.0. CO;2-1.

Dor, A., Burke, T., & Whittington, R. (2007). Assessing the effects of federal pediatric drug safety policies. *The George Washington University Medical Center Newsletter*, 1−16.

Dorfman, L., Wallack, L., & Woodruff, K. (2005). More than a message: Framing public health advocacy to change corporate practices. *Health Education and Behavior, 32*(3), 320−336. Available from https://doi.org/10.1177/1090198105275046.

Drugs.com. *FDA pregnancy risk categories prior to 2015.* (2018). From <https://www.drugs. com/pregnancy-categories.html> Retrieved 17.10.18.

Eckenwiler, L. A., Ells, C., Feinholz, D., & Schoenfeld, T. (2008). Hopes for Helsinki: Reconsidering vulnerability. *Journal of Medical Ethics, 34*(10), 765−766. Available from https://doi.org/10.1136/jme.2007.023481.

Edgar, H., & Rothman, D. J. (1990). New rules for new drugs: The challenge of AIDS to the regulatory process. *Milbank Quarterly, 68*, 111−114.

Eisenberg, R. S. (2005). The problem of new uses. *Yale Journal of Health Policy, Law, & Ethics, 5*, 717−718.

Emanuel, E. J., & Emanuel, L. L. (1992). Four models of the physician-patient relationship. *Journal of the American Medical Association, 267*(16), 2221−2226. Available from https:// doi.org/10.1001/jama.1992.03480160079038.

Entman, R. M. (1993). Framing: Toward clarification of a fractured paradigm. *Journal of Communication, 43*(4), 52. Available from https://doi.org/10.1111/j.1460-2466.1993. tb01304.x.

Fabrizio, C. S. (2011). *Physician's perceptions of the Hong Kong Cervical Screening Program: Implications for improving cervical health* (Unpublished doctoral dissertation). Chapel Hill, NC: University of North Carolina. Retrieved from <https://cdr.lib.unc.edu/record/uui-d:831aabd1-e0d2-4dbb-a89c-934a603b618a>.

Farquhar, C., Armstrong, S., Kim, B., Masson, V., & Sadler, L. (2015). Under-reporting of maternal and perinatal adverse events in New Zealand. *BMJ Open, 5*, e007970. Available from https://doi.org/10.1136/bmjopen-2015-007970.

FDA's Guidance Document To-Do List. (2011). *The pink sheet.*

Feibus, K., & Goldkind, S. F. (2011). Pregnant women and clinical trials: Scientific, regulatory, and ethical considerations. In *Oral presentation at the pregnancy and prescription medication use symposium.* Silver Springs, MD.

Feibus, K. B. (2010). Pregnant women and clinical trials. In *Oral presentation at National Institutes of Health Office of Research on Women's Health workshop: Clinical research enrolling pregnant women.* Retrieved from <https://www.fda.gov/downloads/ScienceResearch/SpecialTopics/WomensHealthResearch/UCM243540.ppt>.

Fieser, J. (1996). Do businesses have moral obligations beyond what the law requires? *Journal of Business Ethics, 15*, 457−468. Available from https://doi.org/10.1007/BF00380365.

Frank, E., & Novick, D. M. (2003). Beyond the question of placebo controls: Ethical issues in psychopharmacological drug studies. *Psychopharmacology, 171*(1), 19−26. Available from https://doi.org/10.1007/s00213-003-1477-z.

Frederiksen, M. C. (2008). Commentary: A needed information source. *Clinical Pharmacology & Therapeutics, 83*(1), 22−23. Available from https://doi.org/10.1038/sj.clpt.6100438.

Freeman, R. E., & Gilbert, D. R. (1992). Business, ethics and society: A critical agenda. *Business & Society, 31*(1), 9−17. Available from https://doi.org/10.1177/000765039203100102.

Frieden, J. (2018, April 9). Pregnancy not a bar to trial participation, FDA says. *MedPage Today.* Retrieved from <https://www.medpagetoday.com/obgyn/pregnancy/72229>.

Gilligan, C. (1982). *In a different voice: Psychological development and women's development.* Cambridge, MA: Harvard University Press in Feminist Ethics.

Gilligan, C. (1987). Moral orientation and moral development. In E. F. Kittay, & D. T. Meyers (Eds.), *Women and moral theory.* Lanham, MD: Rowman & Littlefield Publishers, Inc. (1989).

Gillon, R. (1985). "Primum non nocere" and the principle of non-maleficence. *British Medical Journal, 291*, 130. Available from https://doi.org/10.1136/bmj.291.6488.130.

Glassman, A., & Buse, K. (2008). Politics and public health policy reform. In K. Hehhenhougen, & S. Quah (Eds.), *International encyclopedia of public health* (Vol. 5, pp. 163−170). San Diego, CA: Academic Press.

GlaxoSmithKlein. *Safety and immunogenicity of a trivalent Group B Streptococcus vaccine in healthy pregnant women.* (2014). From <https://clinicaltrials.gov/ct2/show/NCT02046148> Retrieved 17.10.18.

Glover, D. D., Amonkar, M., Rybeck, B. F., & Tracy, T. S. (2003). Prescription, over-the-counter, and herbal medicine use in a rural obstetric population. *American Journal of Obstetrics & Gynecology, 188*(4), 1039−1045.

Goldfarb, N. (2006). The two dimensions of subject vulnerability. *Journal of Clinical Research Best Practices, 2*(8), 1−3.

Goodrum, L. A., Hankins, G. D. V., Jermain, D., & Chanaud, C. M. (2003). Conference report: Complex clinical, legal, and ethical issues of pregnant and postpartum women as subjects in clinical trials. *Journal of Women's Health, 12*(9), 864. Available from https://doi.org/10.1089/154099903770948087.

Greenwood, K. (2010). The mysteries of pregnancy: The role of law in solving the problem of unknown but knowable maternal-fetal medication risk. *University of Cincinnati Law Review, 79*, 267−322.

Hall, J. K. (1995). Exclusion of pregnant women from research protocols: Unethical and illegal. *IRB: Ethics and Human Research, 17*(2), 1−3. Available from https://doi.org/10.2307/3563527.

Harris, L. H. (2003). The status of pregnant women and fetuses in U.S. criminal law. *Journal of the American Medical Association, 289*(13), 1697−1699. Available from https://doi.org/10.1001/jama.289.13.1697.

Health Resources and Services Administration. *National vaccine injury compensation program.* (2018). Retrieved from <https://www.hrsa.gov/vaccine-compensation/index.html>.

Henry, L. M., Larkin, M. E., & Pike, E. R. (2015). Just compensation: A no-fault proposal for research-related injuries. *Journal of Law and the Biosciences*, *2*(3), 645−668. Available from https://doi.org/10.1093/jlb/lsv034.

Henshaw, S. K. (1998). Unintended pregnancy in the United States. *Family Planning Perspectives*, *30*, 24−29. Available from https://doi.org/10.1001/jama.289.13.1697.

Iber, F. L., Riley, W. A., & Murray, P. J. (1987). *Conducting clinical trials*. New York: Plenum Press, In Merton, V. (1993). The exclusion of pregnant, pregnable, and once-pregnable people (a.k.a. women) from biomedical research. *American Journal of Law & Medicine*, *19*(4), 369−451.

Institute for Patient Access (2018, April 13). FDA opens door to pregnant women in clinical trials. *IFPA's Patient Advocacy Blog*. Retrieved from <http://allianceforpatientaccess.org/fda-opens-door-to-pregnant-women-in-clinical-trials/>.

Institute of Medicine Committee on Women's Health Research. (2010). *Women's health research: Progress, pitfalls, and promise*. Washington, DC: National Academies Press. Retrieved from www.nap.edu.

Legal considerations. In Institute of Medicine (U.S.) Committee on Ethical and Legal Issues Relating to the Inclusion of Women in Clinical Studies, A. C. Mastroianni, R. Faden, & D. Federman (Eds.), *Women and health research: Ethical and legal issues of including women in clinical studies: Volume I*. Washington, DC: National Academies Press. Chapter 6, Retrieved from https://www.ncbi.nlm.nih.gov/books/NBK236532/.

Irani, E., & Richmond, T. S. (2015). Reasons for and reservations about research participation in acutely injured adults. *Journal of Nursing Scholarship*, *47*(2), 161−169. Available from https://doi.org/10.1111/jnu.12120.

Jaggar, A. M. (1992). Feminist ethics. In L. Becker, & C. Becker (Eds.), *Encyclopedia of ethics*. New York: Garland Press. cited in Tong, R., & Williams, N. (2018 Winter). Feminist ethics. In E. N. Zalta (Ed.), *The Stanford encyclopedia of philosophy*, https://plato.stanford.edu/archives/win2018/entries/feminism-ethics.

John, P. (2003). Is there life after policy streams, advocacy coalitions, and punctuations: Using evolutionary theory to explain policy change. *Policy Studies Journal*, *32*(4), 488. Available from https://doi.org/10.1111/1541-0072.00039.

Kaposy, C., & Baylis, F. (2011). The common rule, pregnant women, and research: No need to "rescue" that which should be revised. *The American Journal of Bioethics*, *11*(5), 60−62. Available from https://doi.org/10.1080/15265161.2011.560360.

Kass, N. E., Taylor, H. A., & Anderson, J. (2000). Treatment of human immunodeficiency virus during pregnancy: The shift from an exclusive focus on fetal protection to a more balanced approach. *American Journal of Obstetrics and Gynecology*, *182*(4), 1−5.

Kass, N. E., Taylor, H. A., & King, P. A. (1996). Harms of excluding pregnant women from clinical research: The case of HIV-infected pregnant women. *Journal of Law, Medicine & Ethics*, *24*, 36−46. Available from https://doi.org/10.1111/j.1748-720X.1996.tb01831.x.

Kessler, D. A., Merkatz, R. B., & Temple, R. (1993). Author's response to Caschetta MB et al. correspondence: FDA policy on women in drug trials. *New England Journal of Medicine*, *329*(24), 1815−1816. Available from https://doi.org/10.1056/NEJM199312093292414.

Kim, J. H., & Scialli, A. R. (2011). Thalidomide: The tragedy of birth defects and the effective treatment of disease. *Toxicological Sciences*, *122*(1), 1−6. Available from https://doi.org/10.1093/toxsci/kfr088.

Kingdon, J. (1995). *Agendas, alternatives, and public policies* (2nd ed.). New York: Harper Collins College.

Kingdon, J. (2005). The reality of public policy making. In M. Danis, C. Clancy, & H. R. Churchill (Eds.), *Ethical dimensions of health policy*. New York: Oxford University Press, Chapter 6.

Knowles, M. (2018, April 11). FDA pushes to recruit pregnant women in clinical trials. *Becker's Hospital Review*. <https://www.beckershospitalreview.com/quality/fda-pushes-to-recruit-pregnant-women-in-clinical-trials.html>.

Kochanek, K. D., Murphy, S. L., Xu, J., & Arias, E. (2017). Mortality in the United States, 2016. In *NCHS data brief no. 293*. Retrieved from <https://www.cdc.gov/nchs/products/databriefs/db293.htm>.

Koren, G., Bologa, M., Long, D., Feldman, Y., & Shear, N. H. (1989). Perception of teratogenic risk by pregnant women exposed to drugs and chemicals during the first trimester. *American Journal of Obstetrics & Gynecology, 160*, 1190–1194.

Koren, G., & Levichek, Z. (2002). The teratogenicity of drugs for nausea and vomiting of pregnancy: Perceived versus true risk. *American Journal of Obstetrics & Gynecology, 186*(5), S248–S252. Available from https://doi.org/10.1067/mob.2002.122601.

Koren, G., Pastuszak, A., & Ito, S. (1998). Drugs in pregnancy. *New England Journal of Medicine, 338*(16), 1128–1137. Available from https://doi.org/10.1056/NEJM199804163381607.

Koski, E. G. (2005). Renegotiating the grand bargain: Balancing prices, profits, peoples, and principles. In M. A. Santoro, & T. M. Gorrie (Eds.), *Ethics and the pharmaceutical industry*. New York: Cambridge University Press.

Kukla, R. (2005). Conscientious autonomy: Displacing decisions in health care. *Hastings Center Report, 35*(2), 34–44. Available from https://doi.org/10.1353/hcr.2005.0025.

Kukla, R. (2007). How do patients know? *Hastings Center Report, 37*(5), 27–35. Available from https://doi.org/10.1353/hcr.2007.0074.

Langley, G. C., & Egan, A. (2012). The ethics of care in biomedical research committees. *Journal of Clinical Research and Bioethics, 3*, 128. Available from https://doi.org/10.4172/2155-9627.1000128.

Largent, E. A., Joffe, S., & Miller, F. G. (2011). Can research and care be ethically integrated? *Hastings Center Report, 41*(4), 37–46. Available from https://doi.org/10.1002/j.1552-146X.2011.tb00123.x.

Levine, R. J. (1997). *Ethics and regulation of clinical research*. New Haven, CT: Yale University Press, cited in Weijer, C., Dickens, B., & Meslin, E. M. (1997). Bioethics for clinicians: 10. Research ethics. *Canadian Medical Association Journal, 156*(8), 1153–1157.

Li, J., Eisenstein, E. L., Grabowski, H. G., Reid, E. D., Mangum, B., Schulman, K. A., ... Benjamin, D. K., Jr. (2007). Economic return of clinical trials performed under the pediatric exclusivity program. *Journal of the American Medical Association, 297*(5), 480–487. Available from https://doi.org/10.1001/jama.297.5.480.

Lieberman, J. M. (2002). Three streams and four policy entrepreneurs converge: A policy window opens. *Education and Urban Society, 34*(4), 445. Available from https://doi.org/10.1177/00124502034004003.

Lo, W. Y., & Friedman, J. M. (2002). Teratogenicity of recently introduced medication in human pregnancy. *Obstetrics and Gynecology, 100*, 465–473. Available from https://doi.org/10.1016/S0029-7844(02)02122-1.

London, M. (2008). Leadership and advocacy: Dual roles for corporate social responsibility and social entrepreneurship. *Organizational Dynamics, 37*(4), 313–326. Available from https://doi.org/10.1016/j.orgdyn.2008.07.003.

Lott, J. P. (2005). Module three: Vulnerable/special participant populations. *Developing World Bioethics, 5*(1), 30−53. Available from https://doi.org/10.1111/j.1471-8847.2005.00101.x.

Louie, J. K., Acosta, M., Jamieson, D. J., Honein, M. A., & for the California Pandemic (H1N1) Working Group. (2010). Severe 2009 H1N1 influenza in pregnant and postpartum women in California. *New England Journal of Medicine, 362*(1), 27−35. Available from https://doi.org/10.1056/NEJMoa0910444.

Lupton, M. G. F., & Williams, D. J. (2004). The ethics of research in pregnant women: Is maternal consent sufficient? *British Journal of Obstetrics and Gynecology, 111*(12), 1307−1312. Available from https://doi.org/10.1111/j.1471-0528.2004.00342.x.

Lyerly, A., Little, M. O., & Faden, R. (2012). Perspective: Pregnancy and clinical research. *The Hastings Center Report, 38*(6), 53. Available from https://doi.org/10.1353/hcr.0.0089.

Lyerly, A. D. (Nov. 12, 2010). *Personal communication.*

Lyerly, A. D., Faden, R. R., Harris L., & Little, M. O. (2007). Panel session: The second wave: A moral framework for clinical research with pregnant women. In *American Society for Bioethics and Humanities annual meeting.* Retrieved from <http://asbh.confex.com/asbh/2007/techprogram/P6364.HTM>.

Lyerly, A. D., Little, M. O., & Faden, R. (2008). The second wave: Toward responsible inclusion of pregnant women in research. *International Journal of Feminist Approaches in Bioethics, 1*(2), 5−22. Available from https://doi.org/10.3138/ijfab.1.2.5.

Lyerly, A. D., Little, M. O., & Faden, R. R. (2011). Reframing the framework: Toward fair inclusion of pregnant women as participants in research. *The American Journal of Bioethics, 11*(5), 50−52. Available from https://doi.org/10.1080/15265161.2011.560353.

Macklin, R. (2010). The art of medicine: Enrolling pregnant women in biomedical research. *The Lancet, 375*, 632−633. Available from https://doi.org/10.1016/S0140-6736(10)60257-7.

March of Dimes. *Birth defects and other health conditions.* (2013). Retrieved from <https://www.marchofdimes.org/complications/birth-defects-and-health-conditions.aspx>.

Martin, N., & Montagne, R. *U.S. has the worst rate of maternal deaths in the developed world. National Public Radio/WHYY.* (2017). Retrieved from <https://www.npr.org/2017/05/12/528098789/u-s-has-the-worst-rate-of-maternal-deaths-in-the-developed-world>, citing Global Burden of Disease 2015 Maternal Mortality Collaborators. (2016). Global regional, and national levels of maternal mortality, 1990−2015: A systematic analysis for the Global Burden of Disease Study 2015. *The Lancet, 388*(10053), 1775−1812. https://doi.org/10.1016/S0140-6736(16)31470-2.

Mastroianni, A. C., Faden, R., & Federman, D. (Eds.), (1994). *Women and health research: Ethical and legal issues of including women in clinical studies.* Washington, DC: National Academy Press, Institute of Medicine.

Mattison, D., & Zajicek, A. (2006). Gaps in knowledge in treating pregnant women. *Gender Medicine, 3*(3), 169−182. Available from https://doi.org/10.1016/S1550-8579(06)80205-6.

McCormick, T. R. (2013). *Principles of bioethics. Ethics in medicine.* University of Washington School of Medicine. Retrieved from <https://depts.washington.edu/bioethx/tools/princpl.html>.

McCullough, L. B., & Chervenak, F. A. (1994). *Ethics in obstetrics and gynecology.* New York: Oxford University Press, Inc. Available from https://doi.org/10.1046/j.1469-0705.1995.05060424-2.x.

McCullough, L. B., Coverdale, J. H., & Chervenak, F. A. (2005). A comprehensive ethical framework for responsibly designing and conducting pharmacologic research that involves pregnant women. *American Journal of Obstetrics and Gynecology, 193*, 901−907. Available from https://doi.org/10.1016/j.ajog.2005.06.020.

McIntyre, A. (2009, Fall). Doctrine of double effect. In E. N. Zalta (Ed.), *The Stanford encyclopedia of philosophy.* Retrieved from <http://plato.stanford.edu/archives/fall2009/entries/double-effect/>.

McKee, M. (2005). Challenges to health in Easter Europe and the former Soviet Union: A decade of experience. In W. H. Foege (Ed.), *Global health leadership and management.* San Francisco, CA: Jossey-Bass.

Merkatz, R. (1998). Inclusion of women in clinical trials: A historical overview of scientific, ethical, and legal issues. *Journal of Obstetric, Gynecologic, and Newborn Nursing, 27*(1), 78−84. Available from https://doi.org/10.1111/j.1552-6909.1998.tb02594.x.

Merton, V. (1993). The exclusion of pregnant, pregnable, and once-pregnable people (a.k.a. women) from biomedical research. *American Journal of Law & Medicine, 19*(4), 369−451, citing Iber, F. L., Riley, W. A., & Murray, P. J. (1987). *Conducting clinical trials.* New York: Plenum Press.

Metropolitan Atlanta Congenital Defects Program. (2007). 40th Anniversary edition surveillance report. *Birth Defects Research, Part A: Clinical and Molecular Teratology, 79*(2), 1−120. Retrieved from https://onlinelibrary.wiley.com/toc/15420760/79/2.

Miller, F. G., & Wertheimer, A. (2007). Facing up to paternalism in research ethics. *Hastings Center Report, 37*(3), 24−34. Available from https://www.jstor.org/stable/4625743.

Milne, C. (2011). The case for pediatric exclusivity. *BioPharm International, 24,* 12. Retrieved from http://www.biopharminternational.com/case-pediatric-exclusivity.

Mitchell, A. A., Gilboa, S. M., Werler, M. M., Kelley, K. E., Louik, C., & Hernandez-Diaz, S. (2011). Medication use during pregnancy, with particular focus on prescription drugs: 1976−2008. *American Journal of Obstetrics & Gynecology, 205*(1), 51.e1−51.e8. Available from https://doi.org/10.1016/j.ajog.2011.02.029.

Mohanna, K., & Tunna, K. (1999). Withholding consent to participate in clinical trials: Decisions of pregnant women. *British Journal of Obstetrics and Gynaecology, 106,* 892−897. Available from https://doi.org/10.1111/j.1471-0528.1999.tb08426.x, citing Foster, C. (Ed.). (1997). *Manual for Research Ethics Committees.* London: Centre of Medical Law and Ethics.

Moorcraft, S. Y., Marriott, C., Peckitt, C., Cunningham, D., Chau, I., Starling, N., ... Rao, S. (2016). Patients' willingness to participate in clinical trials and their views on aspects of cancer research: Results of a prospective patient survey. *Trials, 17,* 17. Available from https://doi.org/10.1186/s13063-015-1105-3.

National Cancer Institute. *Diethylstilbestrol (DES) and cancer.* (2011). Retrieved from <https://www.cancer.gov/about-cancer/causes-prevention/risk/hormones/des-fact-sheet>.

National Commission for the Protection of Human Subjects of Biomedical and Behavioral Research. *The Belmont report.* (1979). Retrieved from <https://www.hhs.gov/ohrp/regulations-and-policy/belmont-report/read-the-belmont-report/index.html>.

National Institutes of Health. *The 20th century cures act.* (2016). Retrieved from <https://www.nih.gov/research-training/medical-research-initiatives/cures>.

National Institutes of Health Task Force. *National Institutes of Health Task Force on research specific to pregnant women and lactating women. Report to secretary, health and human services congress.* (2018). Retrieved from <https://www.nichd.nih.gov/sites/default/files/2018-09/PRGLAC_Report.pdf>.

Noddings, N. (1984). *Caring: A feminine approach to ethics and moral education.* Berkeley: University of California Press.

Novavax. *A study to determine the safety and efficacy of the RSV F vaccine to protect infants via maternal immunization.* (2015). From <https://clinicaltrials.gov/ct2/show/NCT02624947?term = Novavax> Retrieved 17.10.18.

Novello, A. C. (1992). *Smoking and health in the Americas*. Atlanta, GA: U.S. Department of Health and Human Services, Public Health Service, Centers for Disease Control, National Center for Chronic Disease Prevention and Health Promotion, Office on Smoking and Health, DHHS Publication No. (CDC) 92-8419.

Obstetric and Pediatric Pharmacology and Therapeutics Branch. *Obstetric-Fetal Pharmacology Research Unit (OPRU) network*. (2017). Retrieved from <https://www.nichd.nih.gov/research/supported/opru_network>.

Oliver, T. R. (2006). The politics of public health policy. *Annual Review of Public Health, 27*, 195−233. Available from https://doi.org/10.1146/annurev.publhealth.25.101802.123126.

Philip, N. M., Shannon, C., & Winikoff, B. (Eds.). (2002). Misoprostol and teratogenicity: Reviewing the evidence. In *Report of a meeting at the Population Council, New York, May 22, 2002*. Retrieved from <http://www.misoprostol.org/downloads/Teratogenicity/miso_terato_review.pdf>.

Phumaphi, J. (2005). Building the next generation of leaders. In W. H. Foege (Ed.), *Global health leadership and management*. San Francisco, CA: Jossey-Bass.

Piper, J. M., Ray, W. A., & Rosa, F. W. (1992). Pregnancy outcome following exposure to angiotensin-converting enzyme inhibitors. *Obstetrics and Gynecology, 80*, 429−432.

Pole, M., Einarson, A., Pairaudeau, N., Einarson, T., & Koren, G. (2000). Drug labeling and risk perceptions of teratogenicity: A survey of pregnant Canadian women and their health professionals. *Journal of Clinical Pharmacology, 40*, 573−577.

Presidential Commission for the Study of Bioethical Issues. (2011). *Moral science: Protecting participants in human subjects research*. Washington, DC. Retrieved from <https://bioethicsarchive.georgetown.edu/pcsbi/sites/default/files/Moral%20Science%20June%202012.pdf>.

Rasmussen, S., Olney, R., Holmes, L., Lin, A., Keppler-Noreuil, K., Moore, C., & the National Birth Defects Prevention Study. (2003). Guidelines for case classification for the National Birth Defects Prevention Study. *Birth Defects Research (Part A), 67*, 193−201.

Redington, L. (2009). *The Orphan Drug Act of 1983: A case study of issue framing and the failure to effect policy change from 1990−1994* (Unpublished doctoral dissertation). Chapel Hill, NC: University of North Carolina. <https://cdr.lib.unc.edu/search/uuid:a012aad2-1ab1-43b2-b5ab-0e14740e5e07?anywhere=Redington>.

Rehman, W., Arfons, L. M., & Lazarus, H. M. (2011). The rise, fall and subsequent triumph of thalidomide: Lessons learned in drug development. *Therapeutic Advances in Hematology, 2* (5), 291−308. Available from https://doi.org/10.1177/2040620711413165.

Resnik, D. B., Parasidis, E., Carroll, K., Evans, J. M., Pike, E. R., & Kissling, G. E. (2014). Research-related injury compensation policies of U.S. research institutions. *IRB, 36*(1), 12−19.

Rhoden, N. K. (1987). Informed consent in obstetrics: Some special problems. *Western New England Law Review, 9*(1), 67−88. Retrieved from https://digitalcommons.law.wne.edu/lawreview/vol9/iss1/6.

Rochefort, D. A., & Cobb, R. W. (1993). Problem definition, agenda access, and policy choice. *Policy Studies Journal, 21*(1), 57. Available from https://doi.org/10.1111/j.1541-0072.1993.tb01453.x.

Rodger, M. A., Makropoulos, D., Walker, M., Keely, E., Karovitch, A., & Wells, P. S. (2003). Participation of pregnant women in clinical trials: Will they participate and why? *American Journal of Perinatology, 20*(2), 69−76. Available from https://doi.org/10.1055/s-2003-38318.

Rodriguez, M. A., & Garcia, R. (2013). First, do no harm: The U.S. sexually transmitted disease experiments in Guatemala. *American Journal of Public Health, 103*(12), 2122−2126. Available from https://doi.org/10.2105/AJPH.2013.301520.

Rodriguez, W., Selen, A., Avant, D., Chaurasia, C., Crescenzi, T., Gieser, G., … Uppoor, R. S. (2008). Improving pediatric dosing through pediatric initiatives: What we have learned. *Pediatrics, 121*(3), 530–539. Available from https://doi.org/10.1542/peds.2007-1529.

Rogers, E. M. (1983). *Diffusion of Innovations* (3rd ed.). New York: Free Press.

Rothenberg, K. (1996). The Institute of Medicine's report on women and health research: Implications for IRBs and the research community. *IRB: A Review of Human Subjects Research, 18*(2), 1–3. Available from https://doi.org/10.2307/3563549.

Shields, K. E., & Lyerly, A. D. (2013). Exclusion of pregnant women from industry-sponsored clinical trials. *Obstetrics & Gynecology, 122*(5), 1077–1081. Available from https://doi.org/10.1097/AOG.0b013e3182a9ca67.

Spencer, A. S., & Dawson, A. (2004). Implications of informed consent for obstetric research. *The Obstetrician & Gynaecologist, 6*, 163–167. Available from https://doi.org/10.1576/toag.6.3.163.26999, citing Doyal, L. (1997). Informed consent in medical research. *British Medical Journal, 314*, 1107–1111.

Stachowiak, S. (2009). Pathways for change: 6 theories about how policy change happens. In *Organizational research services*. Retrieved from <http://nmd.bg/wp-content/uploads/2014/04/TW1_Pathways_for_change_6_theories_about_how_policy_change_happens.pdf>.

The Nuremberg Code. *Trials of war criminals before the Nuremberg Military Tribunals under Control Council Law No. 10, Vol. 2, 181–182. Washington, DC: U.S. Government Printing Office, 1949.* (1947). Retrieved from <https://history.nih.gov/research/downloads/nuremberg.pdf>.

Tong, R., & Williams, N. (2018, Summer). Feminist ethics. In E. Zalta (Ed.), *The Stanford encyclopedia of philosophy*. Retrieved from <https://plato.stanford.edu/archives/sum2018/entries/feminism-ethics/>.

UNAIDS/World Health Organization, Joint United Nations Programme on HIV/AIDS. (2007). *Ethical considerations for biomedical HIV prevention trials: Guidance document.* Geneva: UNAIDS. Retrieved from http://www.unaids.org/sites/default/files/media_asset/jc1399_ethical_considerations_en_0.pdf.

U.S. Central Intelligence Agency. *The world factbook: Country comparison: Infant mortality rate.* (2017). Retrieved from <https://www.cia.gov/library/publications/the-worldfactbook/rankorder/2091rank.html>.

U.S. Department of Health and Human Services. *Code of Federal Regulations, Title 45, Part 46, Subpart B. U.S. Government Printing Office via GPO access, 140–143.* (October 1, 2016). Retrieved from <https://www.gpo.gov/fdsys/pkg/CFR-2016-title45-vol1/pdf/CFR-2016-title45-vol1-part46.pdf>.

U.S. Department of Health and Human Services. *Health information privacy: Public health.* (2017). Retrieved from <https://www.hhs.gov/hipaa/for-professionals/special-topics/public-health/index.html>.

U.S. Department of Health and Human Services. *Task force on research specific to pregnant women and lactating women. Report to secretary, health and human services and congress.* (2018). Retrieved from <https://www.nichd.nih.gov/sites/default/files/2018-09/PRGLAC_Report.pdf>.

U.S. Department of Health and Human Services, Public Health Service, National Institutes of Health, Office of Research on Women's Health. (2011). *Enrolling pregnant women: Issues in clinical research.* Bethesda, MD: National Institutes of Health. Retrieved from https://orwh.od.nih.gov/sites/orwh/files/docs/ORWH-EPW-Report-2010.pdf.

U.S. Food and Drug Administration. *Guidance for industry: Enforcement policy concerning certain prior notice requirements.* (2011). Retrieved from <https://www.fda.gov/RegulatoryInformation/Guidances/ucm261080.htm>.

U.S. Food and Drug Administration. (2014). Content and format of labeling for human prescription drug and biological products: Requirements for pregnancy and lactation labeling. *Federal Register, 79*(233), 72064−72103.

U.S. Food and Drug Administration. *Drug research and children.* (2016). Retrieved from <https://www.fda.gov/Drugs/ResourcesForYou/Consumers/ucm143565.htm>.

U.S. Food and Drug Administration. Fact sheet: FDA good guidance practices. (2017). Retrieved from <https://www.fda.gov/AboutFDA/Transparency/TransparencyInitiative/ucm285282.htm>.

U.S. Food and Drug Administration. *Science & research: Pregnancy registries.* (2018a). Retrieved from <https://www.fda.gov/ScienceResearch/SpecialTopics/WomensHealthResearch/ucm251314.htm>.

U.S. Food and Drug Administration. *Gender studies in product development: Historical overview.* (2018b). Retrieved from <https://www.fda.gov/ScienceResearch/SpecialTopics/WomensHealthResearch/ucm134466.htm>.

U.S. Food and Drug Administration. *Pregnancy research initiatives: Enhancing health for mother and child.* (2018c). Retrieved from: <https://www.fda.gov/ScienceResearch/SpecialTopics/WomensHealthResearch/ucm256927.htm>.

U.S. Food and Drug Administration. *The drug development process, step 3: Clinical research.* (2018d). From <https://www.fda.gov/forpatients/approvals/drugs/ucm405622.htm> Retrieved 17.10.18.

U.S. Legal.com. *National Childhood Vaccine Injury Act law and legal definition.* (n.d.). From <https://definitions.uslegal.com/n/national-childhood-vaccine-injury-act-ncvia/> Retrieved 24.10.18.

U.S. National Library of Medicine. *ClinicalTrials.gov.* (n.d.). From <https://clinicaltrials.gov/> Retrieved 17.10.18.

Vojtek, I., Dieussart, I., Doherty, T. M., Franck, V., Hanssens, L., Miller, J., . . . Vyse, A. (2018). Maternal immunization: Where are we now and how to move forward? *Annals of Medicine, 50*(3), 193−208. Available from https://doi.org/10.1080/07853890.2017.1421320.

Weijer, C. (1999). Selecting subjects for participation in clinical research: One sphere of justice. *Journal of Medical Ethics, 25,* 31−36. Available from https://doi.org/10.1136/jme.25.1.31.

Weijer, C., Dickens, B., & Meslin, E. M. (1997). Bioethics for clinicians: Research ethics. *Canadian Medical Association Journal, 156*(8), 1153−1157.

Wendler, D. (2017, Spring). The ethics of clinical research. In E. N. Zalta (Ed.), *The Stanford encyclopedia of philosophy.* Retrieved from <https://plato.stanford.edu/entries/clinical-research/#WhatClinRese>.

Wikipedia. *Teratology.* (2018). From <https://en.wikipedia.org/wiki/Teratology/> Retrieved 15.10.18.

Wikipedia contributors. (n.d.). Institutional review board. In *Wikipedia, the free encyclopedia.* From <https://en.wikipedia.org/wiki/Institutional_review_board> Retrieved 16.10.18.

Wikipedia contributors. (n.d.). Policy. In *Wikipedia, the free encyclopedia.* From <http://en.wikipedia.org/w/index.php?title = Policy&oldid = 484514172> Retrieved 26.10.18.

Wild, V., & Biller-Andorno, N. (2016). Pregnant women's views about participation in clinical research. In F. Baylis, & A. Ballantyne (Eds.), *Clinical research involving pregnant women.* Philadelphia: Springer, Chapter 7.

William, R. J. *Get to know ClinicalTrials.gov. [PowerPoint® presentation].* (2015). Retrieved from <https://www.fda.gov/downloads/ForConsumers/ByAudience/MinorityHealth/UCM465711.pdf>.

Wing, D. A., Powers, B., & Hickok, D. (2010). U.S. Food and Drug Administration drug approval: Slow advances in obstetric care in the United States. *Obstetrics & Gynecology, 115*(4), 825–833.

World Medical Association. *Declaration of Helsinki.* (2008). Retrieved from <https://www. wma.net/wp-content/uploads/2016/11/DoH-Oct2008.pdf>.

Zajicek, A., & Giacoia, G. P. (2007). Obstetric clinical pharmacology: Coming of age. *Clinical Pharmacology and Therapeutics, 81*(4), 481–482. Available from https://doi.org/10.1038/sj. clpt.6100136.

Index

Note: Page numbers followed by "*f*" and "*t*" refer to figures and tables, respectively.

Printed in the United States
By Bookmasters